Exploring
EARTH AND SPACE SCIENCE

7

MET–ORD

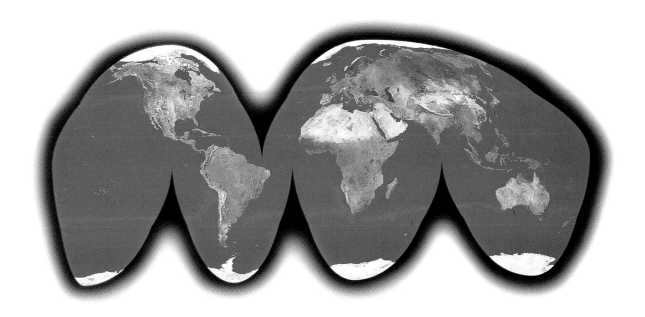

Marshall Cavendish
New York • London • Toronto • Sydney

Marshall Cavendish Corporation
99 White Plains Road
Tarrytown, New York 10591

Website: www.marshallcavendish.com

© 2002 Marshall Cavendish Corporation

Created by **Brown Partworks Limited**

Library of Congress Cataloging-in-Publication Data

Exploring earth and space science.
     p. cm.
     Includes bibliographical references and indexes.
     Contents: 1. Acid and base-Calcium -- 2. Calendar-Continental shelf -- 3. Copper-El
Niño and La Niña -- 4. Energy-Gondwana -- 5. Grassland-Laser -- 6. Light-Meteor -- 7.
Meteorology-Ordovician period -- 8. Ore-Prospecting -- 9. Protein-Star -- 10.
Stratosphere-X ray -- 11. Index.
     ISBN 0-7614-7219-3 (set) -- ISBN 0-7614-7220-7 (v. 1) -- ISBN 0-7614-7221-5 (v. 2)
-- ISBN 0-7614-7222-3 (v. 3) -- ISBN 0-7614-7223-1 (v. 4) -- ISBN 0-7614-7224-X (v.
5) -- ISBN 0-7614-7225-8 (v. 6) -- ISBN 0-7614-7226-6 (v. 7) -- ISBN 0-7614-7227-4
(v. 8) -- ISBN 0-7614-7228-2 (v. 9) -- ISBN 0-7614-7229-0 (v. 10) -- ISBN
0-7614-7230-4 (v. 11)
     1. Earth sciences--Encyclopedias. 2. Space sciences--Encyclopedias. 3.
Astronomy--Encyclopedias

QE5 .E96 2002

550'.3--dc21   00-065801
                                                                                                          CIP
                                                                                                          AC

ISBN 0-7614-7219-3 (set)

ISBN 0-7614-7226-6 (vol. 7)

Printed in Hong Kong

06 05 04 03 02 01 00   5 4 3 2 1

Exploring
EARTH AND SPACE SCIENCE

7

MET–ORD

Marshall Cavendish
New York • London • Toronto • Sydney

Meteorology

The scientific study of changes in Earth's atmosphere, and its use in weather forecasting

Weather happens in Earth's atmosphere, and changes in atmospheric conditions result in changes in Earth's weather and climate. Meteorologists (MEE-tee-uh-RAH-luh-jists) are scientists who observe and study these changes. Some meteorologists use the information they collect to predict what is likely to happen to the weather and climate.

Trying to forecast coming weather has always been of great importance to people. Farmers, for example, need to know when to expect rain to help their crops grow, while fishers need to know when it is too dangerous to set out to sea.

What meteorologists do

Meteorologists make all sorts of observations about the atmosphere, measuring and recording everything from its temperature to wind strength and direction. Temperature, for example, reveals how much heat energy is in the air and the changes that take place as the warmth of the day gives way to the cold of the night.

Temperature affects the air's humidity (the amount of water vapor it contains) because warm air can hold more water vapor. Meteorologists compare the actual amount of water vapor in the air with the maximum air can hold. This comparison is called relative humidity, and it is given as a percentage. If air is saturated (with the maximum water vapor), the relative humidity is 100 percent. So if a reading is 50 percent, the relative humidity is half of what it could be.

HIGHLIGHTS

- Meteorologists observe the changes in Earth's atmosphere, measuring things such as temperature, cloud cover, air pressure, and the movement of the winds. Weather forecasts are based on these observations.

- Although people have observed and attempted to understand the weather since ancient times, scientific meteorology began to develop only during the 17th and 18th centuries.

- Today, the use of computers is essential in making the complex, high-speed calculations that are the basis of modern weather forecasting.

A weather balloon measures atmospheric conditions, while instruments on the ground measure temperature, humidity, pressure, and rain.

The first meteorological satellite, TIROS 1, being launched off Cape Canaveral, Florida, in 1960.

Meteorologists also measure cloud cover—the proportion of the sky that is hidden by clouds. The distance of the clouds above the ground and the different types of cloud are also recorded.

Rain, hail, and snow are ways in which water vapor in the air returns to Earth. Scientists call rain, hail, and snow *precipitation* (prih-SIH-puh-TAY-shuhn). This is also measured and recorded.

Air pressure is the "weight" of the air pressing on each square inch of Earth's surface. It is measured by instruments called barometers (buh-RAHM-uht-uhrz) and is sometimes called barometric pressure. It is often expressed as the number of inches or millimeters of the liquid metal mercury that the atmosphere will support in a tube closed at the top, because this was the arrangement used in the earliest barometers.

Meteorologists measure the direction and speed of winds, and the rate at which these change. The up or down movement of the air is called convection, and measuring this helps to forecast hail, thunderstorms, and tornadoes.

STORY OF SCIENCE

Measuring the Weather

Systematic weather forecasting was introduced in the 19th century by U.S. scientist Joseph Henry (1797–1878). Henry was the first director, from 1846 to 1878, of the Smithsonian Institution in Washington, D.C., where he set up the Smithsonian Meteorological Project. Every day, volunteer telegraph operators across the country sent in records of their weather observations. This became the basis of similar systems throughout the world. In 1873, Henry turned over his network to the U.S. Army Signal Service, which became the National Weather Service. Also in 1873, the International Meteorological Committee was founded, to coordinate the exchange of weather observations by different countries.

The use of maps in weather studies was introduced by German geographer Alexander von Humboldt (1769–1859). It was Humboldt, for example, who came up with isotherms—lines on weather maps that connect areas of equal average temperature– and he used these maps to identify regions of different climate. As well as isotherms, modern weather maps also show isobars—lines connecting areas that are experiencing the same atmospheric pressure.

The third key type of information provided on modern weather maps was introduced in the early 20th century. Norwegian scientist Vilhelm Bjerknes (1862–1951) was the founder of the Bergen Geophysical Institute and its Meteorological Observatory. Together with his son, Jacob Bjerknes (1887–1975), he established a network of weather stations throughout Norway during World War I (1914–1918). Data gathered from these stations led the two to discover that weather conditions are the result of different air masses colliding, and they gave the name "front" to the boundary between two different air masses. Air masses have different characteristics, depending on whether they have crossed an area of Earth that is cold, warm, moist, or dry. How clouds, rain, and winds develop at a front depends on the characteristics of each air mass.

Modern weather maps, published daily throughout the world, show the position of these fronts and their related isotherms and isobars.

Changes in the weather

Although air temperature, humidity, cloud cover, precipitation, air pressure, and wind speed and direction are measured separately, changes in one of them will lead to changes in another. For example, the air temperature beneath clouds will be cooler than it is in cloudless regions, where the air is being warmed by the Sun. As air with a certain amount of water vapor in it cools, its relative humidity goes up, since the cold air cannot hold as much moisture as when it was warm. When the relative humidity reaches 100 percent, it may form a new cloud or add to the cloud above it. A cloud is a suspension of water droplets in the air. When the droplets become too large and heavy to be held up by air, there is precipitation. The air pressure of the warmer air in cloudless regions, on the other hand, may be lower than that under the clouds, so winds may blow there from the air cooled by cloud cover.

Forecasting the weather

Every day, all around the world, thousands of weather observations are made. Some are done by people, while others are done by machines. There are, for example, uncrewed observation stations on remote islands, such as in Alaska's Aleutian group. At fixed times each day, information is collected by solar-powered meteorological instruments and then sent by a small computer to a communications satellite

A satellite image showing Hurricane Andrew near the West Indies in August 1992.

HURRICANE ANDREW
16 - 28 AUG 1992

LOOK CLOSER

Supercomputers

The supercomputers used in central meteorological offices employ mathematical models (formulas) to produce forecasts and can perform two billion calculations a second. The models are based on observations of present weather conditions and predict how these will change in, say, 15 minutes. These forecasts are then used as the starting point for further calculations, which predict the weather for the next few days. Each time, the forecast is approximate, so small errors creep in. The entire set of calculations is restarted several times a day as new observations are collected.

orbiting Earth. From there, the observations are passed on to the U.S. computers of the National Weather Service and shared with the weather monitoring agencies of other countries.

Other observations are made from ships at sea, from survey airplanes, and from weather satellites orbiting Earth. Weather balloons travel high into the atmosphere. Observations are also made by the National Weather Service, at international airports, and smaller airplane landing strips. In many places, at lighthouses for example, amateur weather observers provide important measurements and observations.

In the United States, these observations are sent to supercomputers at the National Meteorological Center near Washington, D.C., and at the National Center for Atmospheric Research in Colorado.

There are Weather Service Forecast Offices in every state, and three each in Texas and Alaska. Local forecasters make further changes to the information, based on their knowledge of weather conditions in their region. Finally, the forecast is made public through radio and television broadcasts and by the private weather companies who deal directly with people in industries such as farming and fishing.

CHECK THESE OUT!
✔AIR PRESSURE ✔ATMOSPHERE ✔CLIMATE
✔CLOUD ✔WEATHER

Microwave

Electromagnetic waves with wavelengths between 1 mm and 30 cm

Microwaves are part of the electromagnetic spectrum, along with radio waves, television waves, visible light, and X rays. With wavelengths between ⅕ inch and 1 foot (1 mm to 30 cm), microwaves have longer wavelengths than X rays and visible light, but shorter wavelengths than television and radio waves.

Using microwaves

Microwaves are vital to many modern technologies, including broadcasting and radar. They are also used in: chemical analysis, since different molecules absorb microwaves with different wavelengths; radio astronomy, since they easily penetrate Earth's atmosphere; and surveying, since reflected microwaves can be used to measure distances very precisely.

Microwaves are the invisible radiation that quickly heats food inside a microwave oven. When the microwaves enter the food, they make the water and other molecules inside the food vibrate more energetically. This generates heat, which cooks the food rapidly from the inside out. In a gas or electric oven, the heat travels slowly from the outside in.

Transmitting microwaves

Unlike light waves, microwaves can pass through fog, clouds, rain, and even smoke, which makes them useful for sending information. Telephone towers, satellites, and television and radio broadcasting stations all use microwaves. The microwaves travel between a pair of antennas (aerials) called a transmitter and a receiver. The transmitter has a device that turns an electrical signal (such as a radio broadcast) into a pattern of microwaves. This pattern is beamed out by the transmitting antenna and picked up some distance away by the receiving antenna, which converts the beam back into an electrical signal.

Radio waves can be transmitted thousands of miles by sending them out toward space and bouncing them back to Earth off a part of the atmosphere called the ionosphere. Microwaves pass through the ionosphere, so they cannot be

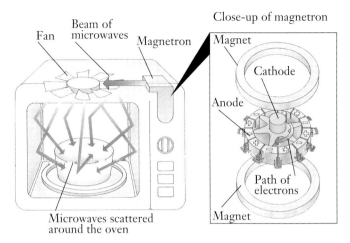

A microwave oven generates invisible microwave radiation that heats food and drinks very rapidly.

transmitted in this way. However, microwave signals can be transmitted from Earth, through the atmosphere, to and from satellites in space.

Radar is a system that detects (picks up) distant objects by sending out a beam of microwaves and calculating how long the beam takes to bounce back. The radar in an airplane, for example, detects obstacles such as another airplane or a mountain range. Weather forecasters use a type of radar that employs the Doppler effect to warn of approaching storms.

Making microwaves

In microwave ovens, the waves are generated in a vacuum tube called a magnetron. This has three main parts—magnets, an anode (positive terminal), and a cathode (negative terminal). The cathode at the center of the tube is heated and gives off electrons, which travel outward to the anode tube. The magnetic field created by the two magnets causes the electrons to swerve on their journey, creating electromagnetic waves. Microwaves can also be generated using a maser—a longer-wavelength version of the laser that produces microwaves instead of visible light.

CHECK THESE OUT!
✔ASTRONOMY ✔DOPPLER EFFECT
✔ELECTROMAGNETIC SPECTRUM
✔ELECTROMAGNETISM ✔ELECTRON ✔IONOSPHERE

Midnight Sun

The period of continuous daylight during summer in the Arctic and Antarctic

In midsummer, near the North and South Poles, there are days when the Sun never sinks below the horizon. These are the lands of the midnight sun. In midwinter, the Sun never rises and the days are dark. This continuous daylight and darkness happen for the same reason that Earth has seasons—because our planet is tilted on its axis (the imaginary line around which it spins) as it orbits the Sun.

Why the Sun never sets

Earth is tilted at 23 degrees and 30 minutes. For this reason, the Northern and Southern Hemispheres both spend six months angled toward the Sun (spring and summer), and six months angled away from it (autumn and winter). Areas in the hemisphere tilted toward the Sun have longer days and shorter nights.

At high latitudes (areas far north or south of the equator), the tilt of Earth's axis is steep and summer days are very long. However, it is only in the areas north of the Arctic Circle (latitude 66 degrees, 33 minutes north) and south of the Antarctic Circle (latitude 66 degrees, 33 minutes south) that the Earth's tilt is steep enough to cause the midnight sun. Within these areas, there is at least one 24-hour period of darkness in winter and at least one 24-hour period of daylight in summer. At the Poles themselves, the day lasts for the whole six months of summer, and night lasts all winter.

Effects of the midnight sun

Animals and birds have patterns of behavior to cope with the polar seasons. Polar bears burrow into the snow to hibernate for the long, dark Arctic winter. Birds that breed high in the polar regions migrate to lower, warmer latitudes in winter. Some animals, such as the Arctic fox, grow white coats in winter to replace their brown summer coats. This helps them to keep hidden from predators or prey in the snowy landscape.

Scientists have used the polar seasons to study the effects of daylight on the human body. The long, dark winter seems to make people sluggish, while continuous daylight can energize people.

A mountain peak lit by the midnight sun at Port Lockroy in the Antarctic.

CHECK THESE OUT!
✔CALENDAR ✔SEASON

Milky Way

Huge spiral galaxy that contains the Solar System and more than 100 billion stars

A view of the constellation Sagittarius, which lies between Earth and the center of the Milky Way.

The Milky Way is Earth's home galaxy. It is an enormous spiral containing all the stars that can be seen with the naked eye from Earth. The spiral forms a disk 100,000 light-years across, with an edge 1,000 light-years thick. A light-year is the distance light travels in empty space in one year, about 5.8 trillion miles (9.4 trillion km).

The Milky Way is so named because it appears as a faint white band, like a path, across the sky on summer nights. This band is made up of millions of stars. A person looking at the Milky Way from Earth is looking out across the disk and so sees a band of stars. A person looking away from the Milky Way is looking out of the galactic plane, into the halo, where stars are more thinly scattered. Earth's Solar System is located about half way out from the center of the galaxy. When people look at the constellation Sagittarius, where the Milky Way is brightest, they are looking toward the center of the galaxy.

Structure of the galaxy

Astronomers know from observing other galaxies and comparing them to measurements from inside the Milky Way that Earth's galaxy has a spiral shape. Seen from the edge, it is thin, with a large ball of stars at its center. Astronomers think there are four main spiral arms swirling out from the center. Named after the constellations within them, they are the Sagittarius Arm, the Perseus Arm, the Orion arm, and the Cygnus arm.

The spiral arms are rich in gas, dust, and hot newly born blue stars called Population I stars. As the Milky Way rotates once every 200 million years, a wave of compression rolls around the disk, pushing gas and dust clouds together and triggering star formation.

The center is very different, with less gas and dust in the central bulge. The stars there, called Population II stars, are older and do not include large blue stars. These are the oldest stars in the galaxy. The Milky Way rotates fastest close to the center. Everything moves around the center of the galaxy, 10 light-years across. It is impossible to see into this region because of the dense star clouds of Sagittarius. However, infrared and radio waves reveal large clusters of very hot stars, and a strong radio source called Sagittarius A*. Astronomers think that Sagittarius A* is a black hole because of its effect on stars around it.

The Milky Way's halo is the ball-shaped shell around the galaxy, containing scattered stars and globular star clusters—large balls of old, red Population II stars—slowly orbiting the brighter regions of the galaxy.

CHECK THESE OUT!
✔ASTRONOMY ✔BLACK HOLE ✔CONSTELLATION
✔EARTH ✔GALAXY ✔STAR

Mineralogy

The scientific study of minerals

A mineral is a natural, usually solid substance that always has the same chemical composition (make-up) and whose atoms form some sort of crystal. Unlike coal and oil, which are organic (contain carbon and hydrogen), minerals are inorganic. Scientists who study minerals are called mineralogists. Mineralogy involves studying mineral properties, classifying families and individual types of minerals, and finding ways to identify them.

Nearly 3,000 minerals are known on Earth. Some are very rare, and only a few, including types of pyroxene (py-RAHK-SEEN), amphibole (AM-fuh-BOHL), and feldspar, plus quartz, olivine (AH-luh-VEEN), and calcite (KAL-SYT), are common. The color and crystal shape of some minerals, such as diamond, are valued for their beauty. Others, such as magnetite, are valuable raw materials for industry, in this case to make iron and steel.

Minerals and rocks

The rocks that make up Earth's crust are composed of minerals. The igneous rock (rock formed from cooled magma) granite, for example, is made mainly of quartz, feldspar, and mica (MY-kuh). Another igneous rock, basalt (buh-SAWLT), is made of feldspar and pyroxene, with some quartz and olivine.

The sedimentary rock sandstone is made mainly of quartz grains, but some types may contain mica, hematite (HEE-muh-TYT), feldspar, calcite, or limonite (LY-muh-NYT). The minerals that make up metamorphic rocks include garnet, chiastolite (kee-AS-tuh-LYT), quartz, and feldspar.

Mineral properties

To identify a mineral, mineralogists make a list of its properties. Many minerals have a distinctive color that can help to identify

Amazon stone is a type of feldspar crystal.
It is highly valued as a gemstone when polished.

HIGHLIGHTS

◆ A mineral is a natural, solid, inorganic substance.

◆ Rocks are composed of minerals.

◆ Minerals can be grouped according to their chemical composition.

◆ Minerals can be identified by properties such as color, hardness, and the shape of their crystals.

them. For example, malachite (MA-luh-KYT) is bright green. However, some minerals can exist in a variety of different colors. In these cases, the mineral may be identified by its streak (the color of its powder). The streak of a particular mineral is always the same. A mineral's luster (shine) can also help to identify it. This can be described as vitreous (glassy) like quartz; metallic like iron pyrite; resinous like spalerite (SPAH-luhr-yt); waxy like chalcedony (kal-SEH-duhn-ee); or silky like gypsum (JIP-suhm).

Mineral hardness

Hardness is a mineral's resistance to scratching. In 1812, German mineralogist Friedrich Mohs (1773–1839) set up a 10-point scale for mineral hardness. The softest mineral, talc, is number 1 on the scale, and diamond, the hardest, is number 10. Everyday objects can be used to test hardness. A fingernail has a hardness of 2.5, for example, while a steel knife is 5.5. If a mineral can be scratched by a steel knife but not a fingernail, its hardness is between 2.5 and 5.5.

Mineral shape

Minerals can be sorted into seven groups called crystal systems by the shape of their crystals. The crystals of a mineral in the cubic system, for example halite, commonly form cubes. Another group is the hexagonal system, which includes quartz. These minerals have six-sided crystals. As well as large, ordered crystals, minerals also form tiny, chaotic sets of crystals. Copper often forms flat, branchlike shapes, for example, while hematite occurs as rounded masses.

Mineral fracture

All minerals fracture (break), but the pieces they split into may be rough and uneven, or flat and smooth—different minerals behave in different ways. When a mineral cleaves (splits) into shapes with flat, smooth surfaces, this is called cleavage. The same shapes are formed every time a crystal of a particular mineral is split. While this can help to identify some minerals, many minerals share the same cleavage shape. Calcite always cleaves into small, rhombic (kitelike) shapes, while galena (guh-LEE-nuh) cleaves into small cubes.

Mineral Groups

Mineralogists usually group minerals by chemical composition. The most common group is the silicates, which are metals combined with silicon and oxygen. There are hundreds of silicates, making up most of the rock in Earth's crust. Silicates include all minerals in the feldspar group, types of amphibole, such as hornblende, and types of pyroxene, such as augite (AW-jyt). Some uncombined elements are described as minerals because they are found in rocks. Gold, silver, and sulfur belong to this group, called the native elements, because they have the simplest possible chemical composition. Other mineral groups include the halide group, which are metals combined with halogens (such as chlorine and fluorine), and sulfides (metals combined with sulfur).

Specific gravity

Most minerals can be identified by their specific gravity (how much heavier or lighter a mineral is compared to the same volume of water). A specimen of galena will feel much heavier to hold than a specimen of calcite of the same size. Galena has a higher specific gravity than calcite.

Using X rays and microscopes

Mineralogists use X rays and microscopes to identify minerals by their chemical composition. An X-ray photograph of a powdered sample is a very accurate means of identification because it shows each mineral's unique atomic structure.

Devices called petrological microscopes pass light through a thin slice of a mineral from different angles, allowing mineralogists to study how the mineral changes the beam of light. An electron microscope uses electrons rather than light waves and can magnify a mineral sample by one million times. The magnification is stronger than an optical microscope, and the picture is clearer. Electron microprobes direct a narrow beam of electrons onto a tiny grain of a mineral, revealing its chemical composition in detail.

CHECK THIS OUT!

✔ATOM ✔CRYSTAL ✔ELECTRON ✔IGNEOUS ROCK ✔METAMORPHIC ROCK ✔SEDIMENTARY ROCK

Mining

The process of extracting materials from under the ground

Coal, iron, aluminum, salt, gold, silver, and diamonds are just some of the useful or valuable materials that are extracted from Earth by mining. Many of these materials, such as diamonds, silver, and iron, are found in rocks deep underground. Despite the expense and danger involved in underground mining, humans have spent much effort inventing new methods and machinery for extracting these materials from the ground.

Thousands of years ago, men and women were digging for flints to make tools and weapons, and earth pigments to paint their cave walls. This primitive digging was the first attempt to get useful materials from Earth. Later (around 1000 B.C.E) came the mining of copper and tin to make bronze. Other useful materials, such as iron, were discovered as mining developed. In Europe, by the 16th century, mining had become a specialist profession. One of the first published scientific works was *De Re Metallica* (*About Metals*) by German scholar Georg Bauer (1494–1555). The book contains a number of illustrations and descriptions of mining techniques, including ways of pumping out water and using bellows to draw air underground for miners to breathe.

Only by the beginning of the 19th century did the search for coal (to use as fuel) become important, followed eventually by oil. Although gunpowder was known in Europe from the 14th century, it was mainly used in warfare. The use of

HIGHLIGHTS

◆ Humans have been digging for useful materials for thousands of years.

◆ Open-pit mining is used to collect rocks that are on or near Earth's surface.

◆ Many rocks are mined deep underground and brought to the surface through shafts.

◆ Gemstones and precious materials such as gold are sometimes extracted from riverbeds and lakes by alluvial mining.

This huge hole in the ground is an open-pit coal mine in West Virginia.

Mining

explosives to crack open hard rock seams was not introduced until late in the 17th century. Before this, miners had to use their own strength. In the mid-19th century, dynamite replaced gunpowder, and engine-powered drills were used.

Mining methods

The very first mining method was to dig a pit into a deposit (a layer of coal or another material) on or near the surface of the ground. Called surface mining or quarrying (KWOH-ree-ing), this method is still used today.

Much of the iron and copper mined today is extracted by a type of surface mining called open-pit bench mining. The open pit is a giant bowl-shaped crater carved into the ground. The "benches" are the terraces (stepped banks of earth) that form the sides of the pit. From the air, the mine looks like a giant dish, with the benches forming a line spiraling down from the rim to the center. Machines called carving rigs scrape away the surface of the bowl-shaped quarry.

Power shovels follow the carving rigs down the spiral trail to pick up the debris (duh-BREE; pile of loose rocks) carved from the quarry side and scoop it into trucks. They can scoop as much as 20 cubic yards (15 cu m) at a time.

Coal deposits near the surface are mostly extracted by strip mining. Strip mines are wide and shallow, with a flat bottom. Heavy machinery digs into the ground and separates the coal from the waste rock.

Alluvial mining

Many hundreds of years ago, the first prospectors (people

searching for materials to mine) discovered that gold, as well as other precious metals and gemstones, could be found in riverbeds or along the banks of rivers and lakes. Material left behind by rivers and floods is described as alluvial (uh-LOO-vee-uhl).

A gold prospector squatting beside a river and washing riverbed silt in a mining pan is practicing alluvial mining. Swirling the water and silt in the pan washes away the lighter materials, such as clay, and, if the prospector is lucky, leaves the heavier gold at the bottom.

Another type of alluvial mining involves using a sluice (SLOOS) box—a long, narrow, wooden box with angled slats nailed along its bottom. River water is run through the sluice box, and the heavier materials, including the gold dust or gemstones, are trapped behind the wooden slats, and the lighter waste is washed away.

STORY OF SCIENCE

Gold Rushes

Gold has always been valuable. Some ancient peoples treasured it because it did not tarnish (discolor) or decay. It was also rare, and only wealthy people could afford to buy gold. During the 19th century, accidental finds led to gold rushes, when hundreds and thousands of prospectors raced to the newly discovered goldfields to try their luck at finding a fortune.

The first U.S. gold rush followed a discovery on January 24, 1848. A carpenter named James Marshall was working at a mill on the American River, about 40 miles (64 km) from Sacramento, California, when he saw some shiny material in the water. Marshall took his find to his boss, Captain John Sutter, who identified it as gold. Within days of the news getting out, the streams near Sutter's mill were packed with crowds of prospectors.

During the following year, 1849, more than 80,000 gold hunters flooded into California. Called forty-niners after the year, these prospectors came from all over the world, not just the United States. Some traveled by boat to the U.S. East Coast and joined U.S. prospectors as they trekked across the Great Plains and through the Rocky Mountains. Others sailed to Central America and then traveled north through the jungles of Panama and Nicaragua, and right across Mexico. Yet others braved the six-month voyage around Cape Horn.

A few prospectors made their fortunes, but most were unlucky. There were other ways to make money from the gold rushes, however. So many people flooded to the goldfields that towns grew up almost overnight—in just three years, for example, the tiny coastal village of San Francisco developed into a city of more than 25,000 people. These new towns and cities filled with people were ideal places to set up successful businesses such as smithies, shops, or ship-building yards.

A large nugget of gold like the one above is a very rare find. Most gold is found as tiny grains of dust.

Underground mining

When deposits of coal and other materials lie far below Earth's surface, underground mining is the only way to get at them. Deep shafts (well-like holes) are drilled, and galleries (sideways tunnels) are driven out from them. To stop the tunnels from collapsing, they are supported by beams and props. Although in the past these were wooden, steel is generally used today.

Below ground, rails are laid for carriages that transport the miners, heavy equipment, and the extracted materials. In some mines, there may be mechanical scrapers or conveyor belts to carry away the extracted rock and waste.

For much of its history, underground mining was done by men, women, children, and ponies. Today, huge drilling machines are used in areas such as the United States. These can mine as much as 15 tons (13.6 tonnes) per minute.

CHECK THESE OUT!
✔CLAY ✔COAL ✔GOLD ✔IRON AND STEEL
✔METALLURGY ✔PROSPECTING ✔SILVER

In large-scale industrial alluvial mining, huge, bargelike machines called dredgers are used on rivers and lakes to scoop up silt and wash it through giant sieves. Gold, diamonds, and tin are often mined using this method.

Mirage

A trick of the light created by light rays passing through air of different temperatures

Thirsty travelers crossing a desert are sometimes fooled by the sight of a shimmering pool of water ahead of them. As they struggle toward it, however, the water seems to move farther away or to disappear altogether. This is because the water is not really there at all. It is an optical illusion (trick of the light) called a mirage (muh-RAHZH). Similar mirages sometimes appear on long, straight roads in hot, sunny weather. Water seems to appear in the middle of the road but vanishes as the viewer gets closer.

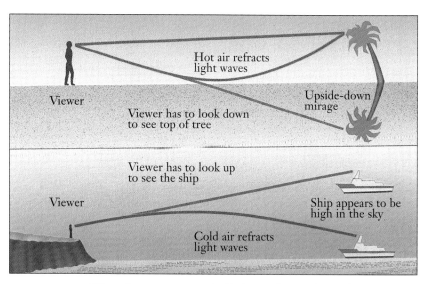

The viewer sees an inferior mirage in a desert (top) and a superior mirage at sea (bottom).

Why mirages happen

Humans see things only when light reflects (bounces) off them and back toward their eyes. Light always travels in straight lines, but it can change speed, for example, when it passes through air of different temperatures. When light rays change speed, they change direction slightly and appear to bend. This bending is called refraction and causes mirages.

The desert travelers' mirage of a pool of water happens because the air near the ground is hotter than the air above it. As rays of light from the Sun pass through colder air into the hotter air near the ground, they refract upward toward the travelers' eyes. The travelers think they are seeing water on the ground, but they are really seeing an image of the sky.

Mirages also make distant objects seem nearer than they really are. Another common desert mirage is when travelers think they see an oasis (oh-AY-suhs), with water and palm trees. The palm trees are real, but they are a long way off, beyond the horizon. They can be seen only because of the refraction of light rays.

Types of mirages

There are two main types of mirages. Inferior (lower) mirages are the most common. They occur when a pool of water or objects around a pool of water are seen. Inferior mirages happen when light rays are refracted upward as they pass from colder air into hotter air.

Superior (higher) mirages are when objects appear to float in the sky. They happen when light rays are refracted downward as they pass from hotter air into colder air. Superior images occur near the North and South Poles, where the air near the icy ground or sea is much colder than the air above. A superior mirage makes objects below the horizon seem closer than they are. A ship may appear to float in the sky, for example, or travelers may think they can see the Sun, when it has really set below the horizon.

A third type of mirage is called the Fata Morgana, which is rare. It is produced when both types of refraction occur together, where a layer of still, warm air hangs over cooler water. An object such as a cliff or a cottage seems to be elongated (stretched out) and can look like a fabulous castle floating half in the sea and half in the air. Such effects are sometimes seen over the Great Lakes in North America and in the Strait of Messina, Sicily.

CHECK THESE OUT!
✔LIGHT ✔OPTICS

Mississippian Period

**The time period that spans
360 to 320 million years ago**

The Mississippian period began about 360 million years ago and lasted 40 million years. Geologists in North America recognize the Mississippian and the Pennsylvanian periods after it (320 to 286 million years ago) as distinct periods of geological time. In the rest of the world, the term Carboniferous period is used to cover the Mississippian and Pennsylvanian.

The name Mississippian comes from the Mississippi valley region, where limestone rocks are rich in fossilized sea animals that lived during the Mississippian period. Mississippian limestone originally formed in a shallow sea. As the period went on, however, the rocks were pushed up to create mountains in what is now eastern North America. Over time, the mountain peaks were eroded (worn away) into mud and clay, which were swept back into the sea. Here, the mud formed thick layers on the sea floor. Added to this were the chalky shells of many millions of tiny dead sea creatures. These layers of sediment were squeezed over many millions of years until it became the limestone typical of the

Mississippian period. When other dead animals sank to the bottom of the sea and were covered by the mud, their bodies became fossilized.

North American Mississippian fossils

The early part of the Mississippian period is sometimes called the age of crinoids, because so many fossils of these creatures have been discovered. Crinoids still live in some oceans today. They look rather like plants and some are called sea lilies, although they are animals with relatives such as sea urchins and starfish. Crinoids commonly live on coral reefs. Sea lilies attach themselves to the seabed by a slender, stalklike body, topped by many arms covered in flowerlike structures. In the center of the arms is the crinoid's mouth, which sifts the seawater for food. Other marine (sea) fossils found in Mississippian limestone rocks include mollusks, brachiopods (shelled creatures that are related to flatworms), and fish.

Marine life

Crinoids and many kinds of mollusks were among the sea animals that thrived during the Mississippian. Gastropods (such as snails), bivalves (such as clams and oysters), and ammonites (spiral-shelled animals) were mollusk groups that also flourished.

Fish had become an important life-form in seas and freshwater lakes during the Devonian, the period before the Mississippian. They continued to thrive, and sharks were among the many fish groups that developed during the Mississippian. Some of the heavily armored Devonian fish died out, however, possibly because they were too heavy to adapt to an increasingly competitive world. The numbers of marine trilobites and brachiopods also dwindled during the Mississippian.

HIGHLIGHTS

♦ The name Mississippian comes from the Mississippi valley, where many fossilized sea animals that lived during the period were found.

♦ Some early sea animals, such as trilobites and brachiopods, died out during the Mississippian, but others, including sharks, evolved.

♦ Early amphibians evolved that could survive on land but bred in water.

♦ Plants flourished on land, and the first kinds of trees developed.

Life on land

The Mississippian saw huge changes in the development of life on Earth. Before this period, few life-forms had existed outside the sea. Now, trees and other plants became common on land, while insects and their relatives flourished, and early amphibians adapted to life out of water.

The first land plants had developed during the Silurian, the period before the Devonian. Although they were very small and only grew near water, they developed a woody tissue called lignin. This was an important step in plant evolution because lignin gives plants a firm structure that allows them to grow tall. Horsetails, clubmosses, and many kinds of ferns spread widely during the Mississippian, and a new group of plants called the conifers also developed. Conifers are able to reproduce in drier conditions. (Pine trees are modern conifers, although they did not evolve until later.) Tall forests grew, with some trees reaching more than 100 feet (30 m). Seed ferns and giant tree ferns flourished in the shade of the trees.

The Mississippian landscape had tall plants, something never before seen on Earth.

The forests were also favorable places for many of the land animals to live. Dragonflies as big as small birds darted between the tree branches and fern fronds. Other land animals found in and around the forests included spiders, centipedes, and harvestmen.

The early air-breathing vertebrates (animals with backbones) called amphibians had developed from the lobe-finned fish during the Mississippian period. The early amphibians crept out of the freshwater streams. Over time, as they evolved, these creatures grew sturdier bones and lost their fishy scales. Like modern amphibians, however, they were tied to water for breeding, because they still laid eggs without shells.

The Mississippian world

During the Mississippian period, shallow seas in which fossil-rich limestones formed were widespread. Earth's landmasses were moving closer together, however, and the seas would withdraw in the Pennsylvanian. Ice sheets formed across a huge landmass that was where Antarctica, South America, and Africa are today. This landmass joined up with another that covered where Europe and Asia now are and a great single continent, called Pangaea, was created. As the two huge landmasses came together, the ocean narrowed and the ocean currents changed. This made temperatures around the world more extreme; cold regions became colder, and hotter areas became hotter. As the ice sheets around the poles grew, more water was locked up as ice, and the sea levels fell.

CHECK THESE OUT!
✔DEVONIAN PERIOD ✔PALEOZOIC ERA
✔PENNSYLVANIAN PERIOD ✔SILURIAN PERIOD

Mixture

Combination of substances in proportions that can vary

Flavored drinks, food, cosmetics, and paints are all examples of mixtures. In general, the substances in a mixture can be identified by analysis and separated using physical methods such as distillation. Commercial mixtures are usually made according to strict recipes that guarantee consistent properties. Other mixtures are less precise. Homemade bread dough is unlikely to contain the same proportions of flour, salt, yeast, and water every time it is made.

Types of mixtures

Scientists classify mixtures as homogeneous (HOH-muh-JEE-nee-uhs) and heterogeneous (HEH-tuh-ruh-JEE-nee-uhs). A homogeneous mixture has a uniform (always the same) composition throughout. A heterogeneous mixture is one in which the parts that make up the mixture can be separated relatively easily. Solutions of solids in liquids (such as sugar in water), mixtures of gases (for example, air), and mixtures of some liquids are examples of homogeneous mixtures. An insoluble solid mixed with a liquid (sand in water) or water droplets suspended in the air as mist and fog are examples of heterogeneous mixtures.

The mixing process

Gases spread out quickly in a process called diffusion, so two gases will form a homogeneous mixture in a short period of time. Many liquids mix together in a similar way to gases, forming a homogeneous mixture if left long enough for diffusion to even out their composition. Liquids that do not mix in this way, such as oil or water, are described as immiscible (ih-MIH-suh-buhl; unable to be mixed). Solids and immiscible liquids can be made to mix by a variety of techniques that require the input of energy and, sometimes, compounds to bind the components together. Emulsions consist of immiscible oil-

A certain amount of sugar dissolves in water to form a homogeneous mixture called a solution.

HIGHLIGHTS

◆ A mixture consists of two or more substances not joined by chemical bonds.

◆ A mixture with a uniform composition is said to be homogeneous.

◆ Physical techniques can separate the components of mixtures using differences in their physical properties, such as boiling point or density.

EVERYDAY SCIENCE

Measuring Fat in Food Mixtures

Food manufacturers often show information about the fat content of food on the packaging. The most common method of measuring the fat content is called solvent extraction. This is done using an assembly of glassware called a Soxhlet apparatus.

First, a sample of food is liquidized or ground to form a paste. The water is removed from the sample, which is placed in a thimble sealed with glass wool. A hotplate heats a solvent to boiling point in a flask. A typical solvent for this procedure is called petroleum ether. Solvent vapors rise through a sidearm in the flask into a condenser, where they cool and turn to liquid. The drops of solvent fall into the thimble containing the food sample, dissolving the fat. Once the fat solution reaches a certain level, it flows through a sidearm into a flask below. After an hour or so, all the fat in the sample has collected in the lower flask. The solvent can then be boiled off and the fat weighed.

based and water-based liquids that are blended together using some form of mixer and usually an emulsifying compound. Mayonnaise is a water-based emulsion. It is made by slowly pouring oil into a mixture of egg and lemon juice (or vinegar) in a blender. The blades of the blender smash the stream of oil into tiny droplets. Natural emulsifiers in eggs surround these oil droplets with a layer that is attracted to the water-based liquid. This prevents the oil forming into larger droplets and separating out of the mixture. As more oil is added, the emulsion becomes more viscous (VIS-kuhs; thick), so the immiscible components of the emulsion cannot separate into layers.

Pastes are mixtures of liquids and insoluble solids. The solid must be ground finely before the liquid is added. Runny liquids make stable pastes only if there is so little liquid present that

The air is a mixture of gases. Water droplets suspended in this mixture make the air misty. Even more water in the suspension causes thick fog.

it simply coats the surfaces of the fine solid particles. More viscous liquids can be added in greater proportions, since their viscosity prevents the solid settling out as sediment.

Separating mixtures

Separating techniques make use of differences in physical properties between the components of mixtures. Fractional distillation separates liquids that boil at different temperatures. It is used to separate petroleum into gasoline, heating oil, and other substances. A machine called a centrifuge (SEN-truh-FYOOJ) separates substances according to their densities. Medical labs use a small centrifuge to separate red blood cells from plasma (the liquid part of blood). Evaporation and cooling can make some solids crystallize from solutions, for example, salt from seawater (evaporation) and rock candy from sugar (by boiling and cooling). Filtration separates solids from liquids (such as coffee grounds from liquid coffee). Solvent extraction separates one or more compound from a mixture by using differences in solubility (how much solvent is needed to dissolve a substance). Chemists use this technique to separate fat from foods or to extract dye molecules from plants. Some techniques are specific to a particular material. Iron can be separated from mixtures with nonmagnetic substances by use of a magnet.

CHECK THESE OUT!
✔ALLOY ✔CENTRIPETAL AND CENTRIFUGAL FORCE
✔DENSITY ✔EVAPORATION AND BOILING
✔PHYSICAL CHANGE

Molecule

The smallest unit of a compound that possesses the chemical properties of that compound

While all matter consists of atoms (the building blocks of chemical elements), few of those atoms are in an unbound state. Of the 90 or so elements that exist in nature, only six exist as free atoms. These are the noble gases of group 18 of the periodic table. In their natural state, the atoms of all other elements are bound to other atoms of the same element or of different elements.

Covalent bonds

The sort of bonding that holds atoms together in a substance depends on the types of atoms in that substance. In some substances, groups of atoms link together by sharing pairs of electrons between atoms. These groups of atoms are called molecules (MAH-lih-KYOOLZ). The term "covalent bond" describes a pair of electrons that links atoms in this way. Covalent bonding is the usual form of linkage between nonmetal atoms, and between most metals and certain groups of nonmetal atoms.

The covalent bonds within a molecule hold the atoms together strongly. By comparison, the attractions between neighboring molecules are weak, which is why substances composed of simple molecules are often gases at room temperature, or liquids or solids with low melting points. This is because the attractions between molecules are easily broken by heat energy. Many simple molecules are polar, that is, they may have some regions where there is a partial negative charge and other regions where there is a partial positive charge. Such substances

A computer model showing liquid water becoming water vapor. In liquid form, the water molecules (H$_2$O) are held together by weak bonds. When the liquid water is heated, these bonds break, and individual water molecules break away from the rest as a gas—water vapor.

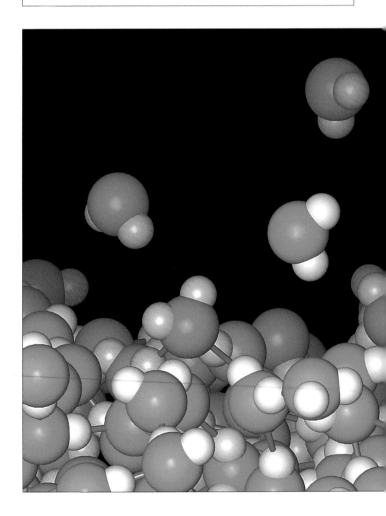

dissolve easily in liquids whose molecules are similarly polar, such as ethanol or water. Other molecules are nonpolar: their charge is evenly balanced through the molecule. These substances dissolve in liquids that are nonpolar, such benzene (C_6H_6).

Macromolecules

Some molecules consist of many thousands or even millions of atoms, all linked together by covalent bonds. These molecules are called macromolecules. Some solid macromolecules, such as glasses, are held together by rigid networks of covalent bonds. These substances tend to be very hard and brittle. They do not melt easily but become soft and stretchy at high temperatures. This softening indicates that heat energy has broken a few of the covalent bonds holding the molecule together.

Plastics are another type of macromolecule, called a polymer. Polymers are formed by linking together in chains many simple molecules, called monomers. The chainlike structure of polymers gives them properties that are very different from those of glasses. The molecules in a polymer can slide past each other, which makes most polymers very flexible. Polymers can be hardened by linking their chains together, forming a three-dimensional network like the structure of a glass.

Macromolecules are found in all living creatures. DNA (deoxyribonucleic acid; a compound that carries inherited information) is a polymer made of two

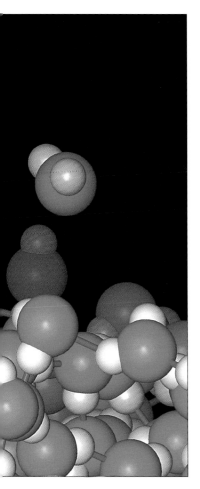

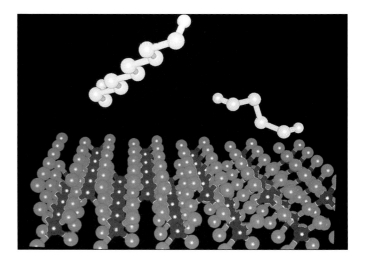

A diagram of a macromolecule of copper oxide (Cu_2O). The oxygen (red) and copper (green) form a network. The yellow molecules are chains of sulfur.

chains wrapped around each other. Proteins are also long-chain polymers, made from 20 different monomers. Proteins fold up into shapes that allow them to perform useful tasks in a living organism. Some living organisms convert sugar molecules into polysaccharides (polymers made of sugar) such as starch and cellulose.

Nonmolecular solids

Metals and salts are examples of solids that are held together by attractions other than covalent bonds. The atoms of a metal each contribute one or more electrons to a pool of electrons that fills the spaces between the atomic nuclei, acting as a "glue" of negative charge that holds the positively charged nuclei together. This type of bonding explains why metals can be drawn out to form wires and hammered to form thin sheets: their atoms easily rearrange by slipping past one another, since the bonding does not hold atoms tightly together. In a salt compound, electrons transfer between atoms of two or more elements to form positive and negative ions. Ions of opposite charge surround each other, making a solid held together by strong attractions.

Valence shells

In order to understand why bonds form between atoms, it helps to understand something of the behavior of electrons in atoms. Each element has a characteristic number of protons in the nuclei of its atoms, and neutral atoms have the same number of electrons as protons. The electrons are strongly attracted to the positive nucleus but repel each other, so they move in patterns that balance the attraction and repulsion. The rules of quantum mechanics (the laws of motion that govern the motion of electrons) allow no more than two electrons to have the same motion. The electrons occupy shells around a nucleus, and each shell has a maximum number of electron

DISCOVERERS

Linus Pauling

In 1931, Linus Pauling (1901–1994) published a paper called *The Nature of the Chemical Bond* in which he explained that many molecular bonds were electron pairs, with one electron coming from each of two atoms. Later he calculated the lengths and angles of bonds and the energies needed to make and break them. In 1954 he won the Nobel Prize for chemistry for this important work.

In the 1940s, Pauling investigated the structure of protein molecules. He observed that many have the shape of a helix. This was very helpful to Rosalind Franklin, Francis Crick, and James Watson in their work on the helical structure of DNA.

In 1962, Linus Pauling (below) won the Nobel Peace Prize for his efforts to stop nuclear testing, which included presenting a petition to the United Nations and publishing a book entitled *No More War*.

pairs, each sharing the same motion that the shell can contain. The shells that are closest to the nucleus fill first and are tightly held by the pull of the positively charged nucleus. These are the core shells. Electrons in the outermost shell, called the valence shell, are less tightly held. The atoms of all elements apart from the noble gases have a valence shell that is only partially full.

Bond formation

The driving force for bond formation is the attraction between electrons and positively charged nuclei. Two unpaired electrons, one from each atom, can form a pair that is attracted to the nuclei on each side. On the other hand, the core electrons are so close to one nucleus that they feel little effect from other nuclei. Further, noble gases do not react because they do not have any unpaired electrons. Covalent bonds form when atoms share a pair of electrons. Chlorine, for example, is one electron short of a full outer shell. If two chlorine atoms form a pair, they can each gain a full outer shell by sharing a pair of electrons. This electron pair is equally attracted to both atoms. Hydrogen atoms are also one electron short of a full shell, and molecules of hydrogen chloride (HCl) gas form by one chlorine atom and one hydrogen atom sharing an electron pair.

Weak bonds

Atoms form molecules by either covelant, ionic, or metallic bonding. However, there are other weaker forces called van der Waals forces that hold liquids and solids together. Named after Dutch physicist Johannes van der Waals (1837–1923), van der Waals forces are caused by attractions between the charged areas of molecules. These charged areas may be permanent, as in polar molecules, or only exist for a very short time due to the random orbiting of electrons. It is these forces that cause gas molecules to group together to form liquids and solids as the temperature drops.

Molecular shapes

While the behavior of electrons in molecules is very complicated, the shapes formed by molecules can be understood by thinking of the bonding pairs of electrons as balloon-shaped pockets joining the atoms. Since these bonds all contain negative charges, they repel one another. This is the basis of the Valence-Shell Electron-Pair Repulsion (VSEPR) model, which compares the bonds around a central atom to party balloons with their necks tied together. Two "balloons" will point directly away from each other, three balloons will form a triangle, and four will form a tetrahedron (pyramid) shape.

Ammonia (NH_3) would seem to contradict VSEPR, since its molecule is the shape of a three-sided pyramid with a nitrogen atom at the top of the pyramid. This is because ammonia has an unbonded electron pair, and that pair forms the fourth point of a tetrahedron. This type of unbonded electron pair is called a lone pair.

Water (H_2O) has a bent shape rather than a straight line shape because the oxygen atom has two electron pairs in addition to the electrons involved in bonding to hydrogen. The four valence electron pairs around the oxygen can be thought of as forming a pyramid with two of the pairs bonded to hydrogen nuclei. In liquid water, the hydrogen atoms of neighboring molecules

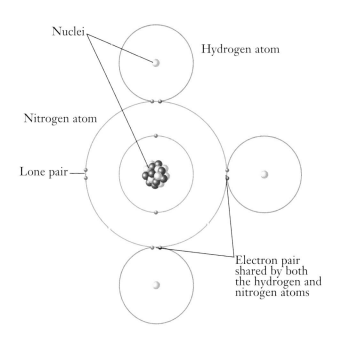

This diagram shows the covalent bonds that hold an ammonia (NH_3) molecule together.

interact with the unpaired electrons to form a weak type of bond called a hydrogen bond. Hydrogen bonding is especially strong in water because the oxygen atom attracts the hydrogen's electron more than most elements. This has important effects: it makes water good at dissolving things, and it increases the boiling point of water so it occurs as a liquid on Earth.

Coordination compounds

The lone pair of ammonia gives its molecules the ability to bond to a hydrogen ion (H^+), which has no electrons. The nitrogen atom continues to have a full valence shell while the hydrogen ion acquires a pair of electrons to fill its valence shell by sharing with nitrogen. All the hydrogen atoms in the resulting ammonium ion (NH^{4+}) are equivalent, and the nitrogen atom carries the positive charge. Boron trichloride (BCl_3) behaves in a similar way: it has two vacancies in its valence shell that it can fill by linking up with the lone pair of ammonia to form a molecule with the formula NH_3BCl_3. The nitrogen atom has a positive charge in this molecule, while the boron atom has a negative charge.

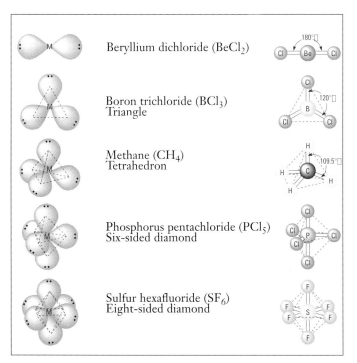

The shapes of certain covalent molecules are shown in the diagram above. Each shape depends on the forces holding the molecule together.

CHECK THESE OUT!
✔CHEMISTRY ✔ELECTRON ✔ION ✔PERIODIC TABLE

Momentum

If a car and a truck are driving side by side at the same speed and they both brake suddenly, the car will stop much more quickly than the truck if the brakes apply the same force. This is an example of momentum, the amount of motion an object has. The truck takes longer to stop with the same braking force because it has more momentum than the car.

Momentum is one of the most important ideas in mechanics, the branch of physics that explains how objects behave when forces act on them. Momentum can be used to explain everything from how a baseball moves when struck to why pedaling faster makes it easier to stay on a bicycle.

How momentum works

Objects that are moving tend to keep moving; those that are still tend to stay still. This tendency is called inertia (i-NUHR-shuh). That objects have inertia is another way of stating the first law of motion worked out by English scientist and mathematician Sir Isaac Newton (1642–1727). Newton showed that it takes a force to make something move (if the object is still) or to make a moving object speed up or slow down.

Momentum explains why this is so. Newton said that an object's linear momentum is equal to the object's mass multiplied by its velocity (speed). Linear momentum also has a direction

—that in which the object is moving. The heavier the object and the faster it is moving, the greater its linear momentum. This is why trucks take longer to stop than cars—they have a greater mass and more momentum. Because momentum is equal to the product of an object's mass and velocity, a car can have the same momentum as a truck only if it is traveling faster. If a truck is ten times heavier than a car, the car must travel ten times faster than the truck to have the same momentum.

Conservation of momentum

The harder a baseball bat is swung, the faster the ball flies away when it is hit. The moving bat has momentum, and some of its momentum is transferred to the ball during the strike. The bat slows down slightly and loses some momentum as the ball is hit. The ball speeds up and gains some momentum. The momentum gained by the

Some of the momentum in the bowler's arm is transferred to the bowling ball. The faster the ball is bowled, the greater its linear momentum.

HIGHLIGHTS

◆ There are two types of momentum: linear momentum (often simply called momentum) and angular momentum.

◆ The linear momentum of a moving object is equal to its mass multiplied by its velocity.

◆ When two objects collide, their total linear momentum before the collision is the same as their total linear momentum afterward. This fact is called the conservation of momentum.

ball is exactly equal to the momentum lost by the bat. This important idea is called the conservation (keeping) of momentum. Whenever two objects collide, the total momentum after the collision is the same as the total momentum before the collision.

Angular momentum

When an object is spinning around an axis (AK-suhs; straight line about which a body rotates), it has an additional kind of momentum, called angular momentum. The angular momentum is equal to the product of an object's rate of spin and its moment of inertia (how the mass is distributed in the object).

Angular momentum also has a direction, which is along the axis of rotation. Linear momentum increases with the object's mass and velocity. However, angular momentum increases both with the speed at which an object is spinning, and with the moment of inertia. If there is no friction acting on the object, the angular momentum will also be lowered.

The moment of inertia increases with the spinning object's mass, and with the average distance of that mass from the object's center of rotation (spin). A flywheel, for example, is a large, heavy wheel that stores energy by

spinning. Most of its mass is concentrated in a heavy rim, some distance from its center of rotation. A spinning flywheel has a large moment of inertia and a great deal of angular momentum. The angular momentum of a flywheel can be increased by applying a torque (TAWRK; turning force) to make it spin faster. The energy stored in its spinning can be used by letting it exert a torque on another object.

Conservation of angular momentum

A spinning object will conserve (keep) its angular momentum by resisting forces that try to change its direction or rate of spin. This is the principle behind the gyroscope (JY-roh-SKOHP). A gyroscope is a wheel mounted so that its axis can turn freely. A gyroscope mounted in an airplane, for example, will point its axis in the same direction, no matter how the airplane turns.

Conservation of angular momentum also explains why it is easier to stay on a speeding bicycle rather than a slow-moving one. The speeding bicycle wheels have a greater angular momentum. They have a greater resistance to forces that try to change their spin—falling off would twist the spinning wheels and change their angular momentum.

CHECK THESE OUT!
✔FORCE ✔MASS ✔MATTER ✔MECHANICS
✔MOTION ✔NEWTONIAN PHYSICS ✔PHYSICS

Monsoon

A strong, regional wind that blows from opposite directions in different seasons

Some of the world's highest rainfall is brought by the Asian summer monsoon.

Monsoons are powerful winds that blow across the tropics and neighboring regions during summer and winter, in opposite directions in each season. The word monsoon is thought to come from the Arabic word *mausim*, which means "seasons." Monsoons have a major influence on the climate in certain areas and affect the lives of people in one-quarter of Earth.

The strongest monsoons occur in southern and eastern Asia. There, the winds blow from the northeast in winter, and from the southwest in summer. Many areas of Earth experience less powerful monsoons, however, including west Africa, where the monsoon shifts from the northeast in winter to the southwest in summer. Northern Australia also has less powerful monsoons, where the winds blow from the southeast during the winter, and from the northwest in summer. The southeastern United States and parts of Central and South America experience slight monsoon winds.

Monsoons are caused by differences in the heating and cooling of air over the land and sea during summer and winter. The temperature of the layer of air nearest Earth's surface comes mainly from the land or sea. In summer, the fierce overhead Sun makes the land hotter than the sea, which makes the air above the land warmer than that above the sea. In winter, however, the Sun's rays are weaker and the days are shorter. The land loses heat much faster than the sea, so in winter, the air above the land is colder than that above the sea. The warmest air lies over the land in summer, while in winter the air is warmest above the sea.

Winds are caused by differences in air pressure and usually blow from areas of high pressure to areas of low pressure. Air pressure, in turn, is affected by air temperature. Warm air rises, leaving an area of low pressure near the surface of the land or sea. Cold air sinks, creating an area of high pressure near the land or sea.

In summer, when the air is warmer over the land, the air pressure is lower over the land, and strong winds blow inshore from the sea, where the air pressure is higher. This is the summer monsoon. In winter, the opposite happens: the monsoon winds blow offshore, from land to sea.

However, monsoon winds do not usually blow straight onshore or offshore. The Earth spins toward the east, which bends winds to the right of their path in the Northern Hemisphere, and to the left in the Southern Hemisphere.

HIGHLIGHTS

- ◆ Monsoons are tropical winds.

- ◆ Monsoons are caused by differences in air temperature between the land and the sea.

- ◆ In summer, monsoons blow in from the sea. In winter, they blow in the opposite direction.

- ◆ Summer monsoons bring heavy rain, which farmers rely on to grow crops. Extreme monsoons might ruin crops and lead to famine.

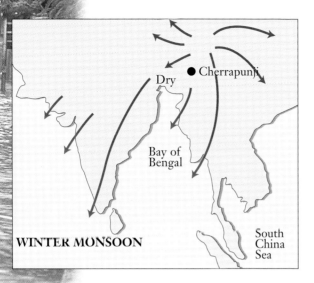
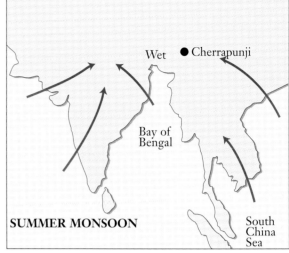

In winter around the Bay of Bengal in southern Asia, the monsoon winds blow offshore. In summer, the monsoons blow onshore, bringing rain and floods.

Monsoon weather

Monsoons greatly affect the local climate. The Asian summer monsoon, for example, brings some of the world's highest rainfall. Cherrapunji, on the southern slopes of the Indian Himalayas, receives an average rainfall of more than 400 inches (10,160 mm) each year. In some years, up to 1,000 inches (25,400 mm) of rain may fall.

Summer monsoons bring rain because they blow in off the sea and are laden with moisture. Towering storm clouds darken the sky and torrential rains fall. In winter, monsoons blow offshore from the land and carry little moisture, so they do not bring heavy rains.

Summer storms occur in groups called monsoon disturbances, each of which lasts a few days. After the monsoon bursts (begins), the disturbances arrive one after another. Separating the disturbances are periods of between 10 and 50 days when the rainfall is lighter.

Monsoons and people

Monsoons are among Earth's most consistent (regular) forms of weather. From year to year, the burst of the monsoons can be forecast almost to the day. Farmers in monsoon areas rely on the heavy summer rains to water their crops, but extreme monsoons do happen, and the results are disastrous. If the rains fail or come too late, the result is a drought (DRAOOT; period of inadequate rainfall), and food crops are ruined.

If, on the other hand, exceptionally heavy rains fall, the rivers swell until they burst their banks and flood the surrounding area. People and animals drown, crops are destroyed, and houses, roads, railroads, and bridges are swept away. The effects can be devastating.

Around the world, floods and drought brought by extreme monsoons are among the causes of famine (FA-muhn; severe shortage of food). For example, in coastal Sahel areas of west Africa bordering the Sahara Desert, droughts caused by the failure of the monsoons from the 1960s to the 1990s resulted in famine.

Forecasting monsoons

Although weather experts can forecast the beginning of monsoons, they cannot predict how long a monsoon will last or how much rain it will bring. Scientists are currently searching for ways to predict the nature of a monsoon about two months before it bursts.

This time would allow farmers to adjust their choice of crops and planting practices to better suit the weather. It would also help lessen the damage caused by extreme monsoons because people could prepare for them.

CHECK THESE OUT!

✔AIR PRESSURE ✔CLIMATE ✔DROUGHT
✔FLOOD ✔RAIN, SLEET, AND SNOW
✔TROPICAL REGION ✔WIND

Moon

**Earth's closest neighbor
and only natural satellite**

The Moon is Earth's natural satellite. It orbits Earth at a distance of about 250,000 miles (400,000 km). Earth's Moon is one of the largest satellites in the Solar System. Its size, and Earth's relatively small size, means the two worlds have important effects on each other, causing tides and eclipses. In describing satellite motion, astronomers distinguish two different periods of motion. The sidereal (sy-DIR-ee-uhl) period is the time it takes the Moon to make one complete orbit of Earth as seen from the distant stars. This time is 27 days, 7 hours, and 43 minutes. Because Earth is moving in its orbit around the Sun at the same time, it takes a little longer for the Moon to get back to the same position with respect to Earth and the Sun. This is the synodic (suh-NAH-dihk) period and is 29½ days. The synodic period is the time between full moons seen from Earth.

The Moon from Earth

Seen from Earth's surface, the Moon is the second brightest object in the sky after the Sun. However, the Moon does not emit its own light but only reflects light from the Sun.

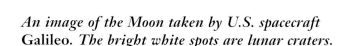
An image of the Moon taken by U.S. spacecraft Galileo. The bright white spots are lunar craters.

The different phases seen throughout the lunar month, as the Moon orbits Earth, are caused by different parts of the Moon being illuminated. At any time, half of the Moon's surface is in light and the other half is in darkness. However, the Moon rotates on its axis (AK-suhs; imaginary line through a spinning body) once each month, so the same side of the Moon always faces Earth. Viewers on Earth see only the part reflecting the sunlight. The phase seen on Earth depends on how much of the illuminated side can be seen.

A lunar month begins at new moon, when the Moon is lined up in the same direction as the Sun in the sky, between Earth and the Sun.

HIGHLIGHTS

♦ The Moon goes through a cycle of phases every month, during which different portions of its illuminated side can be seen.

♦ Occasionally the Sun, Moon, and Earth line up perfectly, causing solar or lunar eclipses.

♦ The Moon's most obvious features are the heavily cratered bright highlands and the dark maria (flat areas people once thought were seas).

Because the near side of the Moon is not in sunlight, it cannot be seen at all from Earth. Over the next few days, as the Moon moves around the sky, it drifts to the east, and a crescent of light appears along one edge. The crescent moon only appears just after sunset, so it can be difficult to see. In good conditions, with the crescent as a clue, the dark side of the Moon is sometimes seen faintly glowing. This is called Earthshine. It happens when the Moon's surface is dimly lit by sunlight reflected from Earth.

Over the first week of a lunar month, the crescent steadily grows (waxes) to a full semicircle, called a first quarter moon, and the Moon starts to set later at night. The next phase, as the semicircle swells to a circle, is called the gibbous (JI-buhs) moon. At 14 days and 18 hours after new moon, the Sun is directly opposite the Moon in the sky, so the Moon is full and is present all night. For the second half of the cycle, as the visible part of the Moon decreases (wanes), it starts to catch up with the Sun again, rising later at night and setting after sunrise.

Eclipses

Earth, the Sun, and the Moon infrequently line up in a completely straight line because the Moon's orbit is tilted. Usually the new moon lies above or below the imaginary line joining the Sun and Earth. Sometimes the three objects line up, however. Because the Sun and Moon appear to be almost exactly the same size in the sky— that is, the Moon, although much smaller than

A lunar eclipse seen from Merritt Island, Florida. The Moon is nearly completely in Earth's shadow.

the Sun, is much nearer to Earth—the Moon can block out the Sun in a solar eclipse.

As the Moon blocks out the bright disk of the Sun, viewers on Earth see the Sun's much fainter outer atmosphere. Solar eclipses happen once every couple of years. They can be seen only from a very small area of Earth's surface where the Moon's shadow falls. Lunar eclipses are much more common—they happen when Earth comes between the Sun and the Moon at a full moon. Because Earth is four times the size of the Moon, its shadow is much larger. It can cover the whole surface of the Moon, and lunar eclipses can be seen from large areas of Earth every year.

Tides

The other major effect the Moon has on Earth is the tides caused by the Moon's gravity pulling on Earth's water and land surfaces. The Moon causes a bulge of a few feet in Earth's seas on the side closest to it, and a lesser bulge of the seas on the opposite side of Earth because the Moon is farther away from this water. As Earth rotates underneath the bulging seas, the tides move around the planet, so the greater bulge is always directly under the Moon.

The tides do not affect only the sea—they also pull on Earth's land surface, causing it to rise and fall a few inches every day. Just as the Moon pulls on Earth, so Earth pulls on the Moon. It is this constant tug of Earth's gravity on the Moon's tidal bulge that has, over billions of years, slowed down the Moon's rotation. It now rotates once in every orbit, with one side always toward Earth.

The lunar surface

The Moon's landscape is divided into two types. The difference between the dark gray patches and the brilliant white surface can be seen from Earth even with the naked eye. Binoculars or telescopes show that the white areas are raised up, rough, and covered in circular pits (craters); these areas are called the lunar highlands.

By contrast, the darker gray areas are low, flat, and have very few craters. Italian astronomer Galileo Galilei (1564–1642), one of the first people to look at the Moon through a telescope, called them *maria* (singular *mare*), which means "seas." The lunar seas contain no water—early space probes to the Moon showed that they are plains of volcanic rock that erupted long ago and filled up the low-lying areas of the surface. Some of the maria have cracks running across them, which were formed when the lava was cooling and solidifying.

The maria have a few craters across their surfaces, but the highlands are covered with them. Before the first space probes, astronomers argued fiercely about whether these craters were the result of volcanoes or impacts from space. The first detailed photographs showed that some craters were just a few millimeters across, so they could not be volcanic. Other craters are also surrounded by material thrown out during the impact, and a few have obvious bright rays

extending for hundreds of miles across the surface. The most impressive crater on the Moon is Copernicus, in the northwest quadrant—it is 58 miles (93 km) across and is surrounded by rays extending for more than 370 miles (600 km).

In 1959, the Russian space probe *Lunik 3* sent back the first pictures of the far side of the Moon. This proved to be very different from the near side. It is more heavily cratered and has only a few very small maria.

The most mysterious regions of the Moon are the poles, which were never photographed during the space probe surveys of the 1960s. In 1994, a probe called *Clementine* found a huge crater near the lunar south pole, probably caused by a comet impact billions of years ago. Because the crater is so deep, it never receives any Sun. Some scientists think that it could have a frozen lake at its base, made from ice dumped by the comet.

The Moon has no atmosphere, so it is completely unprotected from bombardment by meteors (MEE-tee-ORZ; small particles) from space. Therefore, the Moon's surface is heavily cratered. Even the smallest particles, which would burn up in Earth's atmosphere, reach the lunar surface. Although major meteor impacts are now very rare, there are still streams of fine meteor dust drifting around in interplanetary space, pulverizing the surface of the Moon. As a result, the upper part of the Moon's crust is a layer of powdered dust and rock called the lunar regolith (RE-guh-LITH).

The lack of an atmosphere also means that the Moon is exposed to extreme changes in

STORY OF SCIENCE

Missions to the Moon

The Moon was an obvious target for early space missions. It became the center of a space race between the United States and the Soviet Union to see who could put an astronaut on the Moon first.

The first lunar missions were the Soviet Luniks, which crashed into the Moon and sent back pictures from the far side in 1959. The United States caught up in the 1960s, with three series of space probes aimed at paving the way for crewed landings. The Rangers smashed into the Moon's surface while sending back photographs, the Surveyors made soft landings, and the Lunar Orbiters carried out detailed photographic surveys from orbit, looking for possible landing sites. In 1969, the United States put the first of 12 astronauts onto the Moon onboard *Apollo 11*.

The Apollo spacecraft carried out a wide range of experiments. Astronauts set up devices to measure moonquakes and temperature changes, analyze the lunar soil, and return rock samples to Earth. The Soviet Union did not send people to the Moon but explored it with robots instead. By the 1970s, the Moon was forgotten as space probes went farther afield. However, since the 1990s there have been more missions. The *Clementine* and *Lunar Prospector* probes have made maps of the surface and researched possible sites for Moon bases.

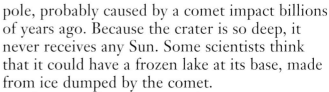

In 1969, U.S. lunar module **Eagle** *landed on the Moon. Its crew were the first to "moonwalk" on the surface.*

The Moon's history

Astronomers are also still arguing about where the Moon came from. Rock samples show that it has great similarities to Earth, but also many differences. At present, the best theory is that Earth formed on its own about 4.6 billion years ago but was struck very early in its history by another large body about the size of Mars. The huge impact vaporized a large chunk of both worlds into orbit around Earth. As Earth cooled down and re-formed, the orbiting material clumped together to form the Moon.

Because the Moon was smaller than Earth, it cooled rapidly, forming a solid crust that soaked up the plentiful meteors in the early Solar System. Eventually, around 3.8 billion years ago, the bombardment rate dropped off suddenly. A few large, late impacts formed deep basins on the Moon, and shortly after this, global volcanoes erupted, flooding the floors of the impact basins with lava, which cooled to form the maria. These volcanoes were probably caused by tidal forces from Earth. The maria are much more common on the Earth-facing side, and the Moon's crust is also thinner on this side. Well over 3 billion years ago, the eruptions stopped as the Moon continued to cool down. The Moon has stayed the same, more or less, ever since.

temperature—made worse because the day and night each last nearly 14¾ days. At noon, the temperatures reach 260°F (127°C), while at night they plummet to –280°F (–173°F).

Lunar activity

The Moon is not a completely dead world. The tides and the constant bombardment by particles from outer space set off tremors below the surface called moonquakes. These moonquakes can sometimes trigger landslides that are visible on the Moon's surface.

The other major form of activity on the Moon is more controversial. Many amateur and professional astronomers have reported seeing brief orange glows on the surface of the Moon. These transient lunar phenomena (TLPs) could be caused by landslips, moonquakes, and impacts releasing trapped gases from below the lunar surface, but the whole issue is still hotly debated.

CHECK THESE OUT!
✔APOLLO MISSION ✔ASTRONOMY
✔ECLIPSE ✔GEMINI MISSION ✔NASA
✔SOLAR SYSTEM ✔SPACE ✔TIDE

Motion

As objects move, they obey precise physical laws. The most important of these are the three laws of motion worked out during the 17th century by English physicist Sir Isaac Newton (1642–1727).

Newton's laws explained exactly why moving objects behave as they do. The laws are usually written down in words, but they can also be put in the form of mathematical equations.

Newton's laws of motion

Newton's first law of motion says that unless a force acts on an object, a still object will stay still and a moving one will keep moving on a linear path (in a straight line). This is another way of describing something called inertia (i-NUHR-shuh)—the tendency of an object to carry on doing what it is already doing until a force makes it behave differently.

Newton's second law explains what happens when a force acts on an object—the object accelerates (speeds up) in the same direction as the force. The bigger the force, the more the

HIGHLIGHTS

- ◆ An object that moves from one place to another obeys Newton's three laws of motion.

- ◆ Newton's laws are used to describe all types of motion. Different forces produce different types of motion.

- ◆ Moving objects have velocity, which means a certain speed in a certain direction.

- ◆ An object's motion can be described using different coordinate systems and may appear to differ according to the system that is used.

object accelerates. However, the greater the object's mass, the less the object accelerates. If someone kicks a football and a tennis ball (which has less mass) with the same force, the tennis ball would travel much farther than the football.

Newton's second law is sometimes written as a mathematical equation: force (F) = mass (m) x acceleration (a).

Newton's third law describes something else that happens when a force acts on an object. This law says that whenever a force (sometimes called an action) acts on an object, it produces another force (called a reaction) that acts in the opposite direction. Newton's third law explains why when someone on a skateboard kicks back against the sidewalk, the skateboard shoots forward. The kick's backward force produces a forward force on the skateboard. Newton's third law also explains how a jet engine pushes

This cyclist exerts a force on the pedals of his bicycle, which makes the wheels go around. The bigger the force used, the faster the bicycle will move.

an aircraft forward by throwing a mass of hot exhaust gas backward, and why someone shoots forward through the water when they pull backward with an oar in a row boat.

Velocity

When a somebody kicks a ball, his or her foot exerts a strong force on it for a very short time. The ball begins to move in the direction of this force with a definite speed, and it gains speed. Together, speed and direction produce something called velocity (vuh-LAH-suh-tee). Speed means moving at a certain rate; velocity means moving at a certain rate and in a certain direction.

A vector (VEK-tuhr) is something that has both a magnitude (MAG-nuh-TOOD; size) and a direction. Velocity, force, acceleration, and momentum are all examples of vectors.

Describing motion

The path of an object's motion can be calculated using a coordinate system (or frame of reference). This is a bit like imaginary graph paper that stretches out horizontally across Earth and upward into space. For the kicked ball, the horizontal direction is the x axis and the vertical

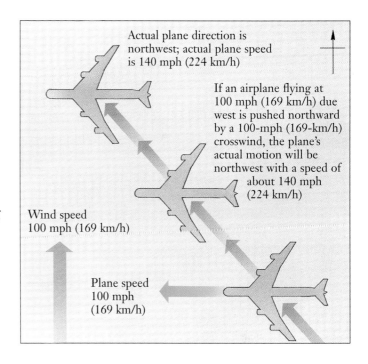

The speed and direction of this airplane produce its velocity, which is a vector.

direction is the y axis. Once the ball has left the person's foot, there are only two forces acting on it—air resistance and Earth's gravity. If the ball is not moving very fast, the air resistance force can be ignored, and the motion becomes very simple. Since gravity acts only in the downward direction, the velocity in the y direction will change, but that in the x direction will remain the same. If the ball was directly over someone's head, that person could run at a constant speed and stay under the ball to catch it. Because there is a force acting downward in the y direction, the velocity of the ball in this direction will decrease at a constant rate as the ball falls to Earth.

To describe the motion of Earth around the Sun, it would make more sense to use a different coordinate system, say one based on the distance between Earth and the Sun and the fraction of a circle that Earth has moved away from one direction in space. Physicists learn how to express Newton's laws in different types of coordinate systems, so the laws can be applied to all types of motion.

CHECK THESE OUT!
✔FORCE ✔GRAVITY ✔MASS ✔MOMENTUM
✔NEWTONIAN PHYSICS ✔RELATIVITY

DISCOVERERS

Aristotle and Galileo

Although Isaac Newton worked out the important laws of motion, he could not have done so without the earlier work of two remarkable scientists—Aristotle and Galileo. Greek philosopher Aristotle (384–322 B.C.E.) believed that there were four elements—air, fire, earth, and water. He also thought that air and fire tended to move upward, while water and earth tended to move downward. Aristotle recognized the crucial importance of forces in the motion of objects.

Many years later, Italian mathematician, physicist, and astronomer Galileo Galilei (1564–1642) put forward important new ideas about motion, including understanding inertia and the way gravity pulls objects to Earth. Newton included the idea of inertia in his laws of motion and then discovered a universal law of gravitation that described the attraction of bodies to Earth, and of all bodies to each other.

Mountain

A mass of rock that rises high above the surrounding landscape

A mountain is a mass of high ground that usually rises steeply to a summit. Definitions of the difference between a hill and a mountain vary, but most geographers agree that any hill that rises 1,800 feet (550 m) above the surrounding area qualifies as a mountain. Few mountains stand alone. Most are found in a row of mountains called a range or chain, which may run for hundreds or even thousands of miles.

Mountains are found on every continent on Earth, and they also rise up from the ocean floor. At 29,028 feet (8,848 m) high, Mount Everest is the highest peak. It lies within the snowcapped Himalayan range in Asia.

The shape of mountains varies greatly. It depends on how they formed, the types of rocks they are composed from, their age, and the forces of erosion that have worn them away. For example, some mountains and whole mountain chains are produced by volcanic activity. Volcanic mountains are often cone-shaped. One example is Mount Fuji in Japan, a dormant (inactive) volcano that stands 12,385 feet (3,776 m) high. Others, such as the 14,690-feet (4,478-m) high Matterhorn in Switzerland, have been worn into a pyramid shape by glaciers. Table Mountain, rising 3,563 feet (1,086 m) above Cape Town in South Africa, is a flat-topped mountain.

Most of the world's major mountain ranges are caused when Earth's crust folds over a long time.

HIGHLIGHTS

- ◆ There are mountains on every continent and on the ocean floor.

- ◆ Most mountains formed as a result of movements of the tectonic plates that make up Earth's crust.

- ◆ There are three main types of mountains: volcanic, fold, and fault-block mountains.

How mountains form

Earth's crust (the surface layer of rock) is made up of a number of huge pieces called tectonic (tek-TAH-nik) plates, which fit together like a massive jigsaw puzzle. The plates may carry either ocean floor or continent, or both. They are not still, but move very slowly over the mantle (the layer below the crust). The plates move apart, collide (hit), or grind against one another. Over time, these movements give rise to various types of mountains. The three main types are volcanic, fold, and fault-block mountains.

Volcanic mountains

Volcanic mountains form when magma (molten rock) wells up from Earth's mantle and erupts as lava through a weak point in the crust. The lava then solidifies to form new rock, which builds up

Magma wells up from Earth's mantle and erupts as lava through a weak point in the crust. Solidifying lava around the volcano crater builds up to form a mountain.

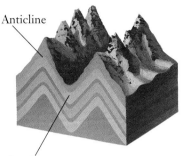

Anticline

Syncline

Colliding tectonic plates slowly crumple the rock into folds, creating mountains and valleys.

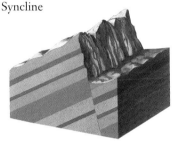

Fault-block mountains form when a huge piece of Earth's crust is forced upward between two faults (cracks) in the crust. A fault occurs where plate movement makes the rocks near the edges of tectonic plates fracture (break).

The formation of different mountains: volcanic (top), fold (middle), and fault-block (bottom).

LOOK CLOSER

Ocean Ranges

The world's longest mountain chain, which includes some of the tallest peaks on Earth, rises from the ocean floor and rarely appears above sea level. Called the midocean ridge, it runs from the Arctic Ocean right along the middle of the Atlantic Ocean. It then splits into two branches, which continue into the Pacific and Indian Oceans.

Most undersea mountains are volcanic and occur along the boundaries between plates or over hot spots. The islands of Hawaii, part of the Pacific branch of the midocean ridge system, are a volcanic chain. Mauna Kea on Hawaii rises 19,500 feet (5,945 m) from the ocean floor and continues for another 13,796 feet (4,205 m) above sea level. It is taller than Mount Everest (29,028 feet or 8,848 m).

around the volcano's crater to form a mountain. Most volcanoes occur along the edges of Earth's plates, where the crust is weak, or at areas in the center of plates called hot spots.

Many volcanoes erupt beneath the oceans where the crust is thin and the edges of a number of plates meet. If a volcano continues to erupt, so much rock may build up that the mountain breaks the water's surface to form an island.

Where two plates collide near the edge of a continent, a range of volcanic mountains might form on land, for example, the Cascade Range in northwestern United States. The range includes the 9,677-feet (2,950-m) high Mount St. Helens, which erupted most recently on May 18, 1980.

Fold mountains

Fold mountains can also form where two plates collide, often at the edge of a continent. The force of the collision slowly crumples the rock into folds. A fold that is forced downward to form a trough is called a syncline. A fold that is forced upward to make an arch is an anticline.

Synclines and anticlines are usually found together and can create mountains and valleys.

Most of the world's major mountain ranges are caused by crustal folding over long periods of time. Fold mountains include the Himalayas of Asia and the Alps in Europe. The Himalayas began to form about 40 million years ago, as the plate bearing India pushed into southern Asia. The Alps formed when the plate bearing Africa pushed north against the plate bearing Europe.

A dome mountain is a particular type of anticline fold that slopes down in all directions from its summit. Dome mountains, such as the Black Hills of Dakota, are formed where pressure from all sides causes uplift.

Fault-block mountains

A fault is a crack in Earth's crust that occurs where plate movement makes the rocks near plate edges fracture (break). Where a chunk of crust is forced upward between two faults, it forms a fault-block mountain, such as the Basin and Range region, which stretches across the western United States north to Idaho.

CHECK THESE OUT!
✔CRATER ✔EROSION ✔FAULT
✔GLACIER ✔PLATE TECTONICS ✔VOLCANO

NASA

U.S. government agency that runs the U.S. program of aeronautical and space research

The National Aeronautics and Space Administration (NASA) is a U.S. government agency. It is in charge of all aspects of space and aeronautic (flight-related) research that are not controlled by the military.

Formation of NASA

NASA's ancestor, the National Advisory Committee for Aeronautics (NACA) was founded in 1915, just 12 years after Orville (1871–1948) and Wilbur (1867–1912) Wright made their first powered flight at Kitty Hawk, North Carolina. NACA was set up to "supervise and direct scientific study into the problems of flight."

NACA did not play a major role in the early days of the space program after World War II (1939–1945). Most of the initial rocket development was carried out by the military. Disputes between different branches of the military soon put the U.S. behind in the race to launch the first satellite. The Russians launched *Sputnik 1* on October 4, 1957. Two months later, the U.S. Navy's Project Vanguard attempted to launch the first American satellite. The rocket exploded on the launchpad and was christened "flopnik." President Dwight Eisenhower (1890–1969) was furious and gave the go-ahead to the U.S. Army's Project Orbiter, which put *Explorer 1* into orbit on January 31, 1958.

NASA space engineers watch the progress of a spacecraft from Mission Control in Houston, Texas.

The United States was losing the space race because there was no single agency controlling its efforts. NASA was founded early in 1958.

NASA's objectives and structure

From the beginning, NASA had a much broader range of interests than NACA. The old agency acted mainly as a research organization, handing on discoveries to the military or manufacturers to put them into use. NASA's main goal was to actively encourage the commercial use of space by pursuing several different paths.

HIGHLIGHTS

◆ NASA was founded in 1958 to help the United States catch up with the Soviet Union (now Russia and other separate states) in the space race.

◆ Spacecraft designed by NASA range from the early one-person Mercury capsules to the space shuttle. NASA has also designed many aircraft.

◆ NASA collaborates widely with other countries and their space agencies. It is currently involved in building the International Space Station.

These included expanding scientific knowledge of Earth, the atmosphere, and space; improving the efficiency and performance of aircraft and spacecraft; and developing vehicles for a crewed space program.

NASA's internal structure shifts as new areas of research arise and old ones finish. However, all of the projects NASA controls fall into one of five branches: space exploration and development, the study of Earth, aeronautics, the study of space, and designing ways of getting into space more easily.

The first branch deals with human exploration of space. Projects within this branch of NASA are focused on increasing knowledge about human spaceflight and ultimately making space exploration a workable commercial enterprise. They include everything from studies of how spaceflight affects humans, to science projects looking at the prospects for efficiently manufacturing certain products in zero gravity.

The second branch of NASA concentrates on improving knowledge of Earth using space-based techniques. NASA's Earth science missions involve remote-sensing satellites, which map Earth's surface, monitor the weather, and study long-term environmental changes.

NASA's aeronautics branch concentrates on developing safe and environmentally friendly new technologies for use in civil and military aircraft. The agency passes the information they learn to the military and businesses.

The space science branch seeks to answer fundamental questions about the nature of the Universe. Its projects range from analyzing meteorites collected on Earth, through funding Earth-based observatories and orbiting telescopes, to sending probes to other planets.

Finally, the space technology project is aimed at reducing the cost of access to space and developing more advanced technologies for

LOOK CLOSER

Galileo

Galileo is one of NASA's most successful robot space probes. It was launched in October 1989 and arrived at its target, the planet Jupiter, in December 1995. Like many NASA space probes, Galileo is a huge international collaboration. It carries experiments built by scientists from Britain, Germany, France, Canada, Taiwan, and Sweden, as well as the United States.

The probe was designed in three segments, each built to study a different part of Jupiter. Several months away from the planet, Galileo split in two. The smaller part—an atmospheric probe—plunged into Jupiter's clouds at a speed of 106,000 miles per hour (171,000 km/h), relaying information to the main probe for sending on to Earth.

The main probe itself had two separate parts. A nonspinning segment carried cameras and other instruments to photograph and measure the surfaces of Jupiter's varied satellites. A larger spinning segment carried instruments to measure the powerful magnetic field around Jupiter itself.

The Galileo probe is launched from a shuttle.

complex future space missions—such as a return to the Moon or a crewed mission to Mars. It is also interested in finding commercial applications for its space technologies.

NASA and astronautics

NASA is best known as the body that organizes crewed spaceflight in the United States. One of its first aims was to put humans into space. NASA called this program Project Mercury.

NASA launched Alan Shepard (1923–1998), the first American astronaut, into space on May 5, 1961, just a month after the Soviet Union had put Yuri Gagarin (1934–1968) in orbit. At the time, U.S. rockets were not powerful enough to put a spacecraft in orbit, so Shepard's flight was just a suborbital hop, climbing briefly into space before falling back to Earth. Nevertheless, just a few weeks later, President John F. Kennedy

The cloud of smoke and flame caused by the explosion of **Challenger** *in 1986.*

(1917–1963) set NASA a seemingly impossible task. He announced his aim to put a U.S. citizen on the Moon before the end of the 1960s.

The buildup to the Moon landings was a golden era for NASA. The agency's budget spiraled to billions of dollars as it raced to develop the technology required to send people to the Moon and bring them back safely. The Mercury program was followed by the Gemini missions, involving two-person capsules that practiced complex maneuvers in space. Meanwhile, engineers were working to design the three-part Apollo spacecraft, with a command and service module that would orbit the Moon and a detachable lunar module to take astronauts down to the surface. They were also developing *Saturn V*, the largest rocket ever.

All this work paid off on July 20, 1969, when Neil Armstrong (born 1930) and Edwin "Buzz" Aldrin (born 1930) became the first astronauts to walk on the Moon. Over the next three years, ten more people followed in their footsteps. However, after the first couple of missions, the public lost interest in Apollo, and NASA found its budget cut back. Faced with tough choices, NASA scrapped the last two Apollo missions in favor of launching *Skylab*, their first space station. Apart from a brief linkup in orbit between U.S. and Soviet astronauts in 1975, this was the end of U.S. crewed spaceflight until 1981. NASA directed nearly all its resources into developing a new spacecraft—the space shuttle.

The space shuttle

The shuttles were built to turn crewed spaceflight into a matter of routine. The spacecraft was designed to be reusable, and it should have cut the costs of reaching orbit drastically. Unfortunately, long delays in the shuttle's development delayed its maiden flight into the 1980s, and the costs never fell as much as predicted.

However, the space shuttles still opened up a new era for spaceflight. NASA has a fleet of shuttles, each able to carry a crew of up to seven. Each shuttle has a large cargo bay that can carry

LOOK CLOSER

The Space Shuttle

NASA has built six space shuttles in all: *Enterprise* (a full-sized prototype), *Columbia, Challenger, Endeavour, Discovery,* and *Atlantis* (the replacement for *Challenger*). Each shuttle system has four separate elements—the shuttle orbiter, which goes into space, a huge liquid fuel tank, and two solid-fuel rocket boosters.

During a shuttle launch, fuel from the liquid fuel tank is burned in the orbiter's main engines, while the solid-fuel rocket boosters provide a short burst of energy to help the shuttle get off the ground. The boosters fall away less than a minute after launch and parachute back into the sea to be recovered and recycled. The orbiter and liquid fuel tank continue to the edge of space, where the last fuel is pumped aboard the orbiter itself and the fuel tank dropped back to burn up in the atmosphere.

The shuttle orbiter, which resembles a large airplane with a 60- by 15-feet (18 by 4.6-m) cargo bay, is able to maneuver (change direction) in orbit using a series of gas thrusters located around the spacecraft. It can deploy or capture and repair satellites, or act as an orbiting laboratory. When the shuttle's mission is over, the orbiter fires its engines to slow it down, and it returns to Earth unpowered, behaving exactly like a giant glider.

The space shuttle **Atlantis** *flies over the deserts of Utah while in orbit around Earth.*

satellites and space probes into orbit or accommodate the European-built laboratory module, Spacelab. The shuttle is able to stay in space for over a week, and so it allowed quite long-duration experiments to be carried out. In some ways, this made up for NASA's lack of a permanent orbiting space station.

Unfortunately, the shuttle program ran into major problems in 1986, when *Challenger* exploded seconds after liftoff, killing all seven crew members. Future launches were put on hold for 18 months until the problems were fixed.

The 1990s were a difficult time for NASA— it suffered a series of failed space probes and constant delays to its plans for a permanent space station in orbit. However, those NASA missions that did succeed, such as the Galileo Jupiter probe and the Hubble Space Telescope, returned spectacular results and did much to repair the agency's image. At the end of the 1990s, NASA's

space station also finally got underway, with the launch of the International Space Station—in partnership with space agencies from Russia, Europe, Japan, Canada, and Brazil.

NASA and aeronautics

Although it is best known for crewed spaceflight, NASA has been widely involved in research. The agency has played a key role in the development of short takeoff and landing aircraft, better engines for airliners, new types of fighter aircraft, and ways to stop pollution and noise from jets. NASA is also researching aerospace planes—craft that fly like jets inside the atmosphere and are propelled like rockets when they reach space.

CHECK THESE OUT!
✔APOLLO MISSION ✔HUBBLE SPACE TELESCOPE
✔LUNAR MISSION ✔MARINER PROBE
✔VIKING PROBE ✔VOYAGER PROBE

National Weather Service

U.S. government agency that monitors the weather and provides forecasts

Will rain fall today, or will it stay dry? Is a storm brewing, or could a tornado be on its way? For answers to these and other questions about the weather, U.S. citizens look to the National Weather Service (NWS). Part of the National Oceanic and Atmospheric Administration (NOAA) of the U.S. Department of Commerce, NWS is the government agency responsible for providing the public with accurate weather information.

In addition to newspapers and television, the NWS also supplies information to universities, aviation (flying) authorities, and a range of private businesses. The NWS also has its own radio broadcasting program, NOAA weather radio. The NWS has its headquarters near

HIGHLIGHTS

◆ The National Weather Service (NWS) provides both short-term weather forecasts and longer-term climate predictions. It issues severe weather warnings covering a wide range of natural hazards.

◆ The NWS uses the latest satellite and radar technology as well as a variety of more traditional monitoring systems to observe and predict changes in the weather.

Tornadoes can be deadly. The NWS tries to predict tornadoes so it can warn people of the danger.

Washington, D.C., and has six regional offices. It also operates the National Severe Storms Forecast Center in Kansas City and the National Hurricane Centers in Miami and Honolulu. The NWS uses nearly 12,000 weather stations to gather climate information.

The NWS provides both short-term weather forecasts and long-term climate predictions. It issues severe weather warnings covering a wide range of hazards, including blizzards, floods, hail, hurricanes, tornadoes, and thunderstorms.

Monitoring the weather

The weather forecasters who work for the NWS receive thousands of meteorological (MEE-tee-uh-RAH-LUH-jih-kuhl; weather science) observations daily, from a combination of volunteer observers, ships, airplanes, automatic weather stations, and weather balloons. Satellites and radar also supply key weather data.

The NWS operates two satellites that observe the United States and surrounding oceans from a height of 22,300 miles (36,000 km) above the equator. The satellites provide information day and night, helping forecasters to keep track of hazards such as hurricanes and flash floods.

The NWS also employs the latest radar technology. Radar is a system that detects (picks up) distant objects by sending out a beam of radio waves and calculating how long the beam takes to bounce back. The NWS's radars can predict a storm's strength, speed, and direction and also help to estimate rainfall and the likelihood of flooding.

The NWS provides the public face of weather forecasting, but behind the scenes the National Centers for Environmental Prediction (NCEP) is at the heart of NWS operations. The NCEP provides information on the weather and conditions at sea to NWS forecasters preparing weather predictions in an area of the United States. The NCEP gets its data from coastal stations and automated buoys floating out at sea.

History of the NWS

In 1996, the NWS was 125 years old. Observing and forecasting the weather have both changed dramatically since 1870, when U.S. President Ulysses S. Grant (1822–1885) authorized the

---

EVERYDAY SCIENCE

Accurate and Inaccurate Forecasts

The weather is not only an important part of our daily news, it often makes the headlines, particularly when a forecast is spectacularly right or wrong. One of the most famous accurate weather forecasts was made on the eve of the Palm Sunday Tornado Outbreak of 1965. The Weather Bureau managed to predict 33 of the 37 tornadoes that swept the Midwest, killing 266 people and injuring 3,261 others. The death and injury toll were still so high because many people ignored the warnings and continued their regular Sunday activities.

Not all predictions are so accurate, though. On September 7, 1900, Texas weather forecasters warned of a hurricane approaching Galveston Island off the coast. However, they failed to predict the tidal surge caused by the hurricane. More than 6,000 people died when a huge wall of storm water washed over the low-lying island. It was the worst natural disaster in U.S. history.

---

setting up of a national weather service. The earliest official weather observers were sergeants in the U.S. Army Signal Service. In 1890, the weather service became a civilian organization called the Weather Bureau. The Weather Bureau became the NWS in 1970.

In the early 20th century, weather forecasts began using more accurate instruments, including kites, to measure temperature, humidity, and winds. In 1960, *Tiros 1*, the world's first weather satellite, was launched.

During the mid-1990s, the NWS modernized and restructured its operations. It did so to take advantage of recent advances in technology. Now the latest communications equipment and superspeed computers help to provide more accurate weather predictions. However, new understanding of weather suggests that exact forecasts may never be possible.

CHECK THIS OUT!

✔CLIMATE ✔FLOOD ✔GLOBAL WARMING ✔HURRICANE ✔METEOROLOGY ✔RADIO WAVE ✔TORNADO ✔WEATHER

Natural Gas

Gaseous hydrocarbon that is used as a fuel to provide energy

Natural gas is a hydrocarbon (compound containing only the elements hydrogen and carbon) found in rocks underground. It often occurs near deposits of crude oil or coal. Natural gas is brought to the surface by drilling. It is used as a fuel for cooking, heating, to generate electricity, and also as a raw material in the industrial production of detergents and plastics.

Methane (CH_4) is the main hydrocarbon in natural gas. Natural gas also contains the hydrocarbons ethane, propane, butane, and pentane, and sometimes small amounts of carbon dioxide, helium, and nitrogen. Methane is the simplest hydrocarbon and has the most hydrogen to its carbon, which allows it to burn smokefree in air. Natural gas is a good fuel because it needs little refining and it supplies clean energy.

How natural gas formed

Like oil, natural gas is a fossil fuel that formed millions of years ago from decaying plankton (microscopic sea organisms). The plankton lived and died in the shallow seas off ancient continents. Their remains sank and were buried in the seabed, where they turned, very slowly, to rock, then oil, and finally to gas.

Today, many of the world's largest oil and natural gas deposits are on land, not under the sea. This is because many oil-rich lands were once covered by ocean. Another reason is that oil and gas migrate (move) under pressure through porous (PAW-ruhs; containing tiny channels) rocks. When they reach an impervious (nonporous) layer, they become trapped in the porous rocks, which are then called "reservoir rocks."

Large-scale rock structures may indicate oil and gas trapped beneath the surface. These include anticlines (the peaks of folds), domes, or faults (fractures) in Earth's crust.

Using natural gas

Natural gas was first extracted on a large scale to be used for fuel during the 1930s. It was named to distinguish it from other gases that can be made from fossil fuels, for example, coal gas.

Oil and gas are extracted by drilling through cap rocks (those above the reservoir rocks). Vast systems of pipelines are built to carry the gas to the area where it will be used. Transporting natural gas is more complicated than transporting oil, which can be carried in barrels and tankers. This is because natural gas ignites much more easily than oil. However, unlike oil, which then needs to be refined, natural gas needs little treatment before it can be used as a fuel.

Natural gas is likely to become an even more important fuel in the future—meeting 90 percent of the world's energy needs, although supplies are limited. Russia and Ukraine hold 33 percent of Earth's known gas reserves, Iran holds 14 percent, and the United States holds 4 percent.

CHECK THESE OUT!
✔COAL ✔COMPOUND ✔ELEMENT ✔ENERGY
✔FOSSIL ✔GAS ✔HYDROCARBON ✔OIL ✔ROCK

Three examples of structures in Earth's crust that can lead to the formation of gas and oil traps.

Fault Trap   **Anticline Trap**   **Salt Dome**

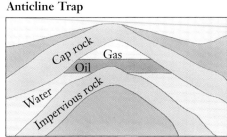

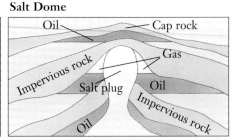

Nebula

A cloud of interstellar gas and dust

The word *nebula* comes from the Latin word for "cloud." For centuries, it was used to refer to any fuzzy patches in the night sky that were not obviously stars or planets. Once the telescope was invented, many of these patches turned out to be clusters of closely packed stars. It is now known that many more patches are other galaxies, so far away that their light takes millions of years to reach Earth.

Today, the word *nebula* is used to describe objects that remain fuzzy no matter how powerful the telescope used. These objects are genuine clouds of gas and dust floating in the space between stars. There are thousands of these clouds throughout Earth's galaxy, the Milky Way. A few clouds can also be seen in other nearby galaxies. There are three main types of nebulas.

Image of the youngest known planetry nebula, Stingray.

Dark nebulas

As their name suggests, dark nebulas absorb light. They can be detected only when they lie in front of a star cluster or another nebula and block out its light. A dark nebula is a cloud of interstellar dust so dense that it absorbs all the light traveling toward it. The best known dark nebula, named for its shape, is called the Horsehead. It lies in the constellation of Orion.

Reflection nebulas

Reflection nebulas have no light of their own but can be seen because they reflect the light of stars toward Earth. They are a mixture of gas and dust, and they are less dense than dark nebulas. As light passes through them, it bounces around, reflecting off the dust particles in a process called scattering. Different colors of light are scattered in different directions, so blue light is reflected out of the nebula at an angle, while red light passes straight through it.

The view of a reflection nebula from Earth depends on its angle to Earth. If someone looks straight through a reflection nebula at the stars behind, the stars appear reddened. If Earth is at just the right angle, a blue reflection nebula can be seen.

Emission nebulas

A nebula that glows with its own light is called an emission nebula. In general, emission nebulas glow because the material inside them is hot enough to shine. However, since nebulas cannot produce their own heat, they all need a power source. Some nebulas glow because they are heated by stars buried deep inside their gas clouds. Emission nebulas are often the material left over when young, hot stars form.

Another type of emission nebula, called a planetary nebula, is a ring-shaped shell of gas blown off by an old star at the very end of its life. The nebula can be lit up by the remains of the star inside—an intensely hot and dense white dwarf—but mostly it simply glows with the remains of the heat it once received in the star.

CHECK THESE OUT!
✔MILKY WAY ✔NOVA ✔PULSAR

Neptune

A cold, blue-green planet rich in helium, hydrogen, water, and methane

The planet Neptune is named for the Roman god of the Sea because of its sea-blue color. Neptune is normally the eighth planet from the Sun, but sometimes Pluto passes closer, on its unusual orbit, and Neptune becomes the ninth. Neptune lies, on average, 2.8 billion miles (4.5 billion km) from the Sun and takes 165 Earth years to complete one orbit.

The planet has several rings and moons and is similar in size and color to Uranus. Otherwise, very little was known about Neptune until 1989, when the *Voyager 2* space probe flew past it.

Appearance and structure

The *Voyager 2* flyby showed that Neptune is a very active world—far more like Jupiter and Saturn than Uranus, which is a quiet and featureless planet. At the time of the *Voyager 2* encounter, Neptune was dominated by a large storm, which was named the Great Dark Spot.

Neptune is the densest of the four "gas giant" planets. The planets in this group (Jupiter, Saturn, Uranus, and Neptune) are composed mainly of matter that would be gaseous in Earth's atmosphere. Astronomers can only guess the nature of Neptune's internal structure.

Neptune pictured from space probe **Voyager 2.** *The* **Great Dark Spot** *in the center is the size of Earth.*

The planet, which is nearly four times the diameter of Earth and has 17 times Earth's mass, might have a molten core of rock and metal surrounded by layers of water ice, methane ice, and liquid ammonia. This "ocean" is surrounded by a thick atmosphere of hydrogen, helium, and water vapor, as well as methane (which makes the planet appear blue-green).

Like Jupiter and Saturn, Neptune generates heat from inside. The planet is probably shrinking very slowly, and the heat and pressure building up around the core as a result cause it to give out roughly twice as much energy as it

HIGHLIGHTS

♦ Neptune is a giant planet, with a diameter nearly four times larger than that of Earth.

♦ Neptune appears blue-green because of the methane in its atmosphere.

♦ Neptune gives out twice as much heat as it receives from the Sun. This heat helps drive the planet's strong winds.

STORY OF SCIENCE

Discovery of Neptune

Neptune was the first planet to be discovered as a result of a deliberate search. Ever since Uranus was discovered in 1781, astronomers had noticed that the planet did not orbit in the way scientists expected. Throughout the early 1800s, the planet ran ahead of its predicted position, and then, in 1829, lagged behind.

Two young astronomers, Frenchman Urbain-Jean-Joseph Le Verrier (1811–1877) and Englishman John Couch Adams (1819–1892), independently unraveled the mystery. They realized that Uranus must be affected by the gravity of another large planet in orbit beyond it. Both Le Verrier and Adams gave precise predictions of the planet's location so that more senior astronomers could search for it. Unfortunately for Adams, the Astronomer Royal in Britain, George Biddel Airy (1801–1892), dismissed his calculations at first, before eventually ordering a brief search in July and August 1846. During this search, British astronomers saw Neptune but did not realize it was the planet they were looking for.

Meanwhile, Le Verrier had approached a more openminded astronomer. German Johann Gottfried Galle (1812–1910), of the Berlin Observatory, received a letter from Le Verrier on September 23, 1846. He found the planet through his telescope that same night, exactly where Le Verrier predicted.

Strangely, Neptune might have been discovered more than 200 years earlier. Italian astronomer Galileo Galilei (1564–1642) saw the planet with one of the earliest telescopes while observing Jupiter but did not recognize it as a planet.

receives from the Sun. It is this heat that drives Neptune's cloud systems. The planet has the highest winds in the Solar System, reaching speeds of up to 1,500 miles per hour (2,400 km/h) and blowing around the planet in the opposite direction from the way the planet spins. Something similar drives the winds on Jupiter and Saturn. Uranus is the only gas giant that does not generate heat. This is probably why it is comparatively quiet.

The Great Dark Spot of Neptune rotates counterclockwise as well as circling the planet at 750 miles per hour (1,200 km/h). *Voyager 2* also pictured a white cloud feature, called the Scooter because it scoots around Neptune at an even faster speed. Astronomers think that the different cloud colors are caused by different chemicals freezing into clouds at different heights in Neptune's atmosphere. The white clouds are probably frozen methane crystals, and the dark areas may be windows into deeper atmosphere.

Rings and moons

Neptune has at least four thin rings made up of millions of dark, dustlike particles. Each particle orbits the planet separately and is kept in line by shepherd satellites. As well as the four complete rings, there are five ring arcs. These are small segments of rings that have either broken up or never formed properly.

Apart from the small shepherd satellites, Neptune has two main moons: Triton (TRY-TAHN) and Nereid (NIR-ee-UHD). Triton is a giant satellite, larger than Earth's Moon, and surprisingly active. It has ice geysers (GY-zuhrz; springs of hot water and steam) and smoking volcanoes. Some parts of Triton's surface look similar to the Moon's, while others are covered with pinkish ice.

Triton may be one of a class of small, icy outer worlds, including Pluto. Triton seems to have formed in the depths of the outer Solar System and was captured by Neptune during a close encounter in the distant past. Neptune's gravity has now forced it into a perfectly circular orbit, but one that goes clockwise around the planet, unlike most large satellites, which revolve in counterclockwise orbits. Nereid, a small moon with a highly irregular orbit, may have been forced into this orbit by the arrival of Triton.

CHECK THESE OUT!
✔ASTRONOMY ✔COSMOLOGY
✔SATELLITE ✔SOLAR SYSTEM ✔VOYAGER PROBE

Neutron

Neutral particle that is present in atomic nuclei and that takes part in nuclear reactions

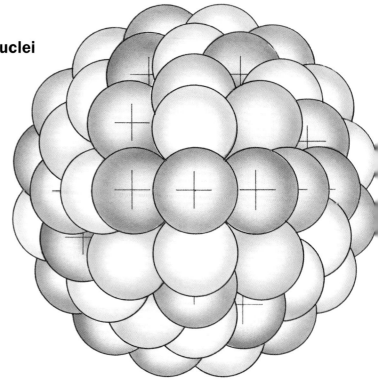

An atomic nucleus showing neutrons in blue and positively charged protons in red.

Neutrons are subatomic particles that have no electrical charge. Together with positively charged protons, they are present in the nuclei of all atoms except the majority of hydrogen atoms. Neutrons are strongly attracted to both protons and other neutrons within atomic nuclei. Without this attraction, the repulsion between positively charged protons would make it impossible for nuclei with more than one proton to exist. Neutrons have no charge but they are magnetic. If a beam of neutrons is fired through a magnetic field, the magnetic force deflects the path of the beam.

While all atoms of a given element have the same number of protons in their nuclei, they can have different numbers of neutrons. For example, iron atoms always have 26 protons, but they can have 28, 30, 31, or 32 neutrons. Types of atoms that only differ by their numbers of neutrons are called isotopes.

Only certain isotopes are stable. Those that are unstable disintegrate by radioactive decay until they form a stable combination. This can occur by fission, the splitting of a nucleus with the release of an additional neutron, by the emission of alpha particles (two protons and two neutrons), by beta decay (the conversion of a neutron into a proton with the release of an electron), or by positron emission (the conversion of a proton into a neutron with the emission of a positron).

Neutron discovery

By the 1930s, scientists had detected electrons and protons. They worked out the existence of a neutral particle from the fact that the atomic masses of elements are greater than they would be if atoms only had electrons and protons.

In 1932, British physicist James Chadwick (1891–1974) devised a method for detecting neutrons. Chadwick bombarded beryllium metal with alpha particles, which produced neutral particles. When the neutral particles were allowed to collide with gas molecules, the gas molecules recoiled as if they had been hit by particles with the same mass as protons.

Nuclear reactions

Italian-born scientist Enrico Fermi (1901–1954) later used Chadwick's method to bombard various substances. In many cases, he found that the target substance became radioactive. Fermi had discovered neutron capture: nuclei absorbing a neutron to form a different isotope of the same element. For example, phosphorus 31 becomes phosphorus 32. Many of the isotopes formed by neutron capture are highly unstable. In one case, uranium 235 splits into two smaller atoms plus several neutrons and releases energy.

CHECK THESE OUT!
✔ATOM ✔ELECTRON ✔FISSION ✔ISOTOPE
✔NUCLEAR PHYSICS ✔PARTICLE PHYSICS
✔PROTON ✔PULSAR ✔RADIOACTIVITY

Newtonian Physics

The branch of physical science based on Newton's laws of motion and gravitation

Sometimes called classical physics, Newtonian physics is an approach to physical sciences based on the discoveries of English physicist Sir Isaac Newton (1642–1727). Newton discovered laws that accurately describe and predict the movements of physical objects and the effects of gravity between different objects.

Until the early 20th century, most people thought Newtonian physics was a perfect description of the way the Universe works. The discoveries of relativity and quantum theory showed that Newton's laws break down in extreme conditions. Relativity affects objects moving at very high speeds, and quantum theory affects objects on a subatomic scale. For the vast majority of everyday situations, however, Newton's laws are still very accurate.

Early discoveries

At the start of Newton's life, the basic laws of physics were still largely undiscovered, but a lot of the groundwork had already been done. In the 1580s, Italian scientist Galileo Galilei (1564–1642) had proved that Earth's gravity had the same effect on all objects, regardless of their

PHILOSOPHIÆ
NATURALIS
PRINCIPIA
MATHEMATICA.

Autore *JS. NEWTON*, *Trin. Coll. Cantab. Soc.* Mathefeos Profeffore *Lucafiano*, & Societatis Regalis Sodali.

IMPRIMATUR·
S. PEPYS, *Reg. Soc.* PRÆSES.
Julii 5. 1686.

LONDINI,

Juffu *Societatis Regiæ* ac Typis *Jofephi Streater*. Proftat apud plures Bibliopolas. *Anno* MDCLXXXVII.

Newton published his theories about motion and gravitation in **Principia Mathematica** *in 1687.*

mass. In other words, a light object and a heavy object of the same shape, dropped from the same height, would take the same time to hit the ground. He also showed that it was not necessary to constantly apply a force to an object in order to keep it moving.

In 1609, German astronomer Johannes Kepler (1571–1630) published the first two of his three laws of planetary motion. Based on detailed observations made by Danish astronomers Tycho

HIGHLIGHTS

◆ Newtonian physics works for all everyday situations. It only breaks down at extremely small scales or at extremely high speeds.

◆ Newton took 20 years to develop his original ideas linking gravity to the movement of the planets into his great work, the *Principia Mathematica*.

◆ Newton's theory was the first one able to explain every aspect of planetary motions and the movement of objects on Earth.

Isaac Newton

Isaac Newton had an extraordinary life. He was born on Christmas Day, 1642, at Woolsthorpe, Lincolnshire, England. He grew up in the care of his grandmother. Although not a good pupil at school, he enjoyed making mechanical devices as a boy.

Although Newton's mother wanted him to be a farmer, his uncle recognized the boy's ability, keeping him in school and allowing him to eventually go to Cambridge University. There, Newton buried himself in private study while taking little interest in his courses. Between 1661 and 1665, he developed the mathematical methods called calculus that would allow him to develop his physical theories.

In 1665, Cambridge University was closed because of disease outbreaks and Newton returned to Woolsthorpe, where his interest in gravitation was first sparked. When he got back to Cambridge in 1667, he concentrated mainly on optics, until his astronomer friend Edmond Halley (1656–1742) reignited his interest in gravity in 1684. It was Halley who eventually paid for the publication of the *Principia Mathematica* in 1687.

After the publication of *Principia*, Newton was a famous man. As well as writing another book called *Opticks*, he became a member of Parliament and was in charge of making English money. He died in 1727 at the age of 85.

(1546–1601) and Sophie (around 1556–1643) Brahe, Kepler's laws were the first accurate description of how the planets move as they orbit around the Sun under the influence of gravity.

By Newton's time, there was plenty of information about how objects moved in relation to each other and under the influence of gravity. However, no one had yet explained exactly how gravity worked. Scientists realized there was a force that pulled objects toward Earth's center, and some thought it was a form of magnetism. No one connected it to the force that kept the planets in their orbits.

Newton's "crucial experiment," in which he used a prism to split light into all its colors.

Newton's laws

The *Principia* changed all this forever. Newton wrote it in just 18 months, but it was the result of 20 years of work. He developed an entire system of mathematics, which he called calculus, to support it. The *Principia* has three separate books, each of which covers an important advance in Newton's theory of the Universe.

In the first book, Newton outlines his three laws of motion:

1. Every body continues in a state of rest or uniform motion in a straight line unless acted on by an outside force.

2. The acceleration of an object matches the force acting on it. The same force will cause half the acceleration on an object of twice the mass.

3. For every action (force in one direction) there is an equal and opposite reaction (a force in the opposite direction).

In the second book, Newton discusses the movement of bodies through gases or liquids. At the time, no one knew that space is a vacuum, and the most popular theory to explain the motion of planets suggested that they were kept in their orbits by swirling whirlpools of tiny particles constantly pushing them around. Newton proved that this theory was incapable of explaining Kepler's laws of planetary motion.

In the third book, Newton put forward his own theory of universal gravitation, summarized by a simple equation giving the force (F) between two objects of masses M and m, separated by a distance R:

$$F = \frac{GMm}{R^2}$$

In the equation, G is a constant, called the gravitational constant, and depends on the units used to measure the other elements of the equation. G is always very small. The equation explains why the force of gravity has a significant effect only when at least one of the two bodies involved has a very large mass. It also explains why the force of gravity diminishes rapidly as the distance between the objects increases.

Newton then went on to show how his laws of motion and gravitation could explain Kepler's laws perfectly, as well as describing other natural events such as tides and the appearance of comets. It is little wonder that he called this third book *The System of the World*.

Newton and light

Newton's work turned the Universe into an elaborate clockwork model. Many scientists loved its elegance and simplicity, but others did not like the idea of a force that acted at a distance through empty space. It took more than 50 years before the theory of vortices was completely abandoned by other scientists.

Meanwhile, Newton moved on to revolutionize the science of optics (lenses, mirrors, and light). Newton had already conducted many experiments with optical instruments and had invented his own form of telescope, the Newtonian reflector. When he was a student, he devised what he called the crucial experiment, splitting light into a spectrum of different colors by passing it through a prism, and then showing that the individual colors of light could not be split further. This proved that the colors were a property of light itself and not of the prism.

Newton believed that light was fundamentally a particle. He developed his theories of light and color in *Opticks*, published in 1704, where he used geometry to explain many aspects of the behavior of light.

Newton's particle theory of light held for over a century, until the early 1800s when scientists finally proved that light behaves somewhat like a wave. Today, it is clear that light has both wave and particle characteristics, but it is impossible to devise an experiment that will show both properties at the same time.

Gravity

Newton realized that if an apple was pulled to Earth by gravity, then the Moon would be too.

Newton and the apple

The most famous story about Isaac Newton is that he discovered gravity by watching an apple fall from a tree in his garden in 1666. This is not exactly true. Newton did not discover gravity—it was already well known. Newton's unique revelation was that gravity extended into the sky and did not simply hold things down on Earth. If gravity could reach the apple, then why should it not reach out as far as the Moon and beyond?

Newton realized that if gravity did extend to the Moon, then gravity might be the force that kept the Moon in orbit around Earth, counterbalancing the Moon's own natural tendency to move in a straight line.

CHECK THESE OUT!
✔COLOR ✔FORCE ✔GRAVITY ✔LIGHT
✔MASS ✔MATTER ✔MECHANICS ✔MOMENTUM
✔PHYSICS ✔QUANTUM THEORY ✔RELATIVITY

Nitrogen

A gaseous element that is exchanged between the atmosphere, Earth's crust, and living organisms

Nitrogen is a gaseous element that makes up about 78 percent of air. It has no color, odor, or taste, and it does not burn. Nitrogen was discovered in the 1770s by three scientists who were working individually: British scientists Joseph Priestley (1733–1804) and Daniel Rutherford (1749–1819), and Swedish chemist Carl Scheele (1742–1786). All three scientists were studying the gas that is left when all the oxygen has been removed from air by burning, which is almost pure nitrogen.

The German name for nitrogen—*stickstoff*—means "suffocating substance." Rutherford noted nitrogen's suffocating property when he discovered that mice could not survive in air that had no oxygen. Similarly, the French word for nitrogen—*azote*—comes from the Greek for "without life."

Physical properties

Nitrogen is the seventh element in the periodic table and the lightest element of group 15. The element nitrogen exists as diatomic molecules (consisting of two atoms, N_2) that form a gas at room temperature. Liquid nitrogen boils at –320.4°F (–195.8°C) and freezes at

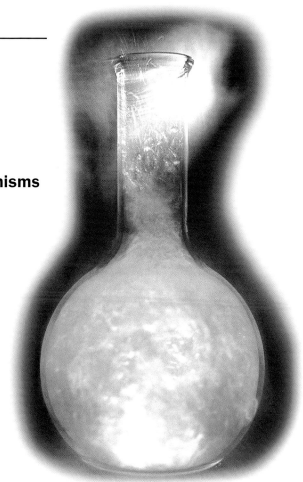

A flask of liquid nitrogen. Liquid nitrogen is used as a coolant in laboratories and industry.

–345.8°F (–209.9°C). It is obtained from air by repeated cycles of compression, cooling, and expansion. The air gets colder with each cycle until it finally turns to liquid. Fractional distillation (when the liquids are boiled off at different temperatures) then separates the parts of liquid air, producing pure nitrogen, oxygen, and argon, with traces of other gases. Liquid nitrogen is widely used to flash-freeze foodstuffs and to freeze and refrigerate tissue samples and cultures of bacteria and viruses.

Chemical properties

The two atoms in a molecule of nitrogen are held together by a strong triple bond. Therefore, nitrogen gas is relatively unreactive—the triple bond between the two nitrogen atoms has to break for nitrogen to react. Compounds of nitrogen with carbon, hydrogen, oxygen, and sometimes sulfur form a wide range of natural and synthetic substances, including proteins, amino acids, nylon-type polymers, and synthetic

HIGHLIGHTS

♦ Nitrogen accounts for 78 percent of the volume of air.

♦ Compounds of nitrogen with elements such as carbon, hydrogen, and oxygen include proteins, nylon-type polymers, synthetic dyes, and ammonia.

♦ Ammonia is an important raw material for the manufacture of fertilizers and other nitrogen compounds.

azo dyes. At temperatures above around 390°F (around 200°C), nitrogen reacts with alkali and alkaline earth metals to form nitrides, which contain the N^{3-} ion.

Ammonia

Ammonia is produced by the bacterial decomposition of organic materials such as plants and animals. For many decades, the main source of ammonia for industry was the breakdown of coal to make gas for heating and lighting, of which ammonia was a by-product.

In 1909, German chemist Fritz Haber (1868–1934) devised a process for making ammonia from hydrogen and nitrogen by passing these two gases over a catalyst of iron oxide. This process was scaled up by German chemist Carl Bosch (1874–1940). The Haber-Bosch process made available huge quantities of ammonia. This was used to develop certain nitrogen compounds that are used in dyes and many medical drugs.

Ammonia is a pungent, irritant gas that is poisonous at low concentrations. It was used in early refrigerators because it can be liquefied by pressure alone and absorbs a great deal of heat as it evaporates. Ammonia is also the basis of many cleaning products.

Nitric acid

Much of the ammonia produced by industry is used to make nitric acid. Nitric acid (HNO_3) is used to produce dyes, explosives, and fertilizers. The Ostwald ammonia oxidation process burns ammonia in air on a platinum gauze at around 1800°F (1000°C) and produces nitric oxide:

$$4NH_3 + 5O_2 \rightarrow 4NO + 6H_2O.$$

The nitric oxide produced by this reaction forms nitric acid with oxygen and water:

$$4NO + 3O_2 + 2H_2O \rightarrow 4HNO_3.$$

CHECK THESE OUT!
✔ATMOSPHERE ✔CHEMICAL REACTION
✔ELEMENT ✔GAS ✔POLLUTION

LOOK CLOSER

Nitrogen Cycle

The nitrogen cycle is a collection of chemical and biochemical reactions that circulates nitrogen among the atmosphere, living organisms, and Earth's crust. A number of natural processes change molecules of nitrogen gas into compounds of nitrogen that can dissolve in groundwater and enter growing plants. Certain types of bacteria in soil change nitrogen from the air into nitrogen compounds in a process called nitrogen fixation. Many of these bacteria thrive in the roots of legumes (LEH-GYOOMZ; beans, peas, and clover). Legumes absorb nitrogen compounds from the soil to form proteins and amino acids. Lightning is another process of nitrogen fixation. Nitrogen combines with oxygen in the air to form nitrogen oxides, which combine with atmospheric moisture to form nitric and nitrous acids, which fall in rain. Nitric acid reacts with minerals in soil to form nitrates, while nitrous acid forms nitrites.

Animals take in nitrogen compounds when they eat plants and other animals. Their bodies use the nitrogen to make many vital substances, such as proteins and DNA. Bacteria then change the nitrogen content of animal urine, feces, and dead bodies into ammonia (ammonification), which enters the soil. Further bacteria change some of this ammonia into nitrites and nitrates (nitrification), which are absorbed by growing plants. Other bacteria change nitrates into nitrogen gas (denitrification), which is released into the air.

Since the late 18th century, human industrial activities have increasingly affected the nitrogen cycle. Fossil fuels, such as coal and oil, contain small amounts of nitrogen. When these fuels burn, nitrogen forms oxides of nitrogen, which contribute to acid rain and smog (smoke and fog). Also, large amounts of synthetic nitrates are now used as fertilizers, but they have been washed into waterways, clogging them with nitrate-loving vegetation.

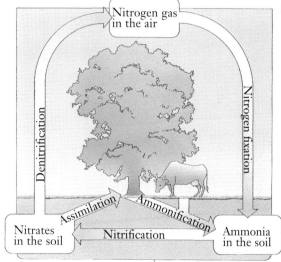

Nitrogen gas in the air · Denitrification · Nitrogen fixation · Assimilation · Ammonification · Nitrates in the soil · Nitrification · Ammonia in the soil

Nonmetal

Elements that do not conduct electricity well and lack a shiny appearance

Most of the 90 naturally occurring substances in the periodic table of chemical elements are either metals or nonmetals. Metals tend to be shiny solids that are good conductors (carriers) of electricity and heat, whereas nonmetals are generally gases and tend to be much poorer conductors. Metalloids are substances that have properties that fall between those of metals and nonmetals.

Scientists classify 17 of the 90 natural elements as nonmetals. The nonmetals are hydrogen (chemical symbol H), helium (He), carbon (C), nitrogen (N), oxygen (O), fluorine (F), neon (Ne), phosphorus (P), sulfur (S), chlorine (Cl), argon (Ar), selenium (Se), bromine (Br), krypton (Kr), iodine (I), xenon (Xe), and radon (Rn).

Properties of nonmetals

The physical properties of metals and nonmetals are very different. Metals tend to be much more dense (have more matter in a given volume) than nonmetals and tend to have higher melting and boiling points. With the exception of mercury,

The nonmetal carbon has three forms. The black powder below is the form called graphite.

all metals are solids at room temperature and normal atmospheric pressure. Of the nonmetals, one is a liquid (bromine), four are solids (carbon, phosphorus, sulfur, and iodine), and the remaining 12 are gases.

The biggest difference between metals and nonmetals is the way they conduct electricity and heat. Metals tend to be good conductors, whereas, with the exception of a form of carbon called graphite and a few of the gases, such as neon, nonmetals are very poor conductors.

Metals tend to have broadly similar chemical properties. The chemical properties of nonmetals are much more varied. Helium, neon, argon, krypton, xenon, and radon form a group called the noble (inert) gases—they are very stable and barely react with other elements at all. Other nonmetals, such as fluorine and oxygen, are very reactive. They can form compounds either by gaining electrons from metals to make what are called ionic bonds, or they can share electrons with other nonmetals to make covalent bonds.

Where nonmetals are found

Together, the light nonmetal gases hydrogen and helium make up more than 99 percent of all the atoms in the Universe. Earth's atmosphere is a mixture of nonmetal gases, including around 78 percent nitrogen and 21 percent oxygen. The remainder is mostly argon, water vapor, and carbon dioxide. Taking into account Earth's crust, oceans, and atmosphere, oxygen is by far the most common element—accounting for almost half of all Earth's mass.

Oxygen is also a common element in living things. Together with hydrogen, nitrogen, carbon, sulfur, and phosphorus, it makes up 97 percent of the human body.

CHECK THESE OUT!
✔COMPOUND ✔ELECTRON ✔ELEMENT
✔METAL ✔METALLOID ✔PERIODIC TABLE

North America

The Northern Hemisphere landmass that is Earth's third largest continent

The continent of North America stretches from the Arctic Ocean in the north to South America in the south. It is bordered by the Pacific Ocean to the west and the Atlantic Ocean to the east. As well as the United States, the continent includes Canada, Mexico, Central America, the islands of the Caribbean Sea, and the huge island of Greenland. This amounts to one-sixth of all the dry land on Earth.

North America is the third largest continent, covering 9,362,000 square miles (15,849,866 sq km). Africa and Asia are larger, but only Asia has more than North America's 37,000 miles (59,540 km) of coastline. North America is a land of contrasts, stretching across four climate regions. The continent covers nearly 5,000 miles (8,040 km)—from land with permanently frozen soil in the far north, to the tropical climate of the south. At 20,320 feet (6,194 m) above sea level, its highest point is Mount McKinley in Alaska. Its lowest is Death Valley, California, at 282 feet (86 m) below sea level.

From the Arctic Ocean in the north, thinning to the Isthmus (IS-muhs; narrow strip of land) of Panama in the south, North America is roughly triangular. In the north, it measures some 5,400 miles (8,690 km) across. In the south, the narrowest point is only 31 miles (50 km) wide.

HIGHLIGHTS

◆ North America includes the United States, Canada, Mexico, and the countries of Central America.

◆ The huge continent has four major climate zones from the Arctic to Central America.

◆ North America reached its present size and shape about 600 million years ago.

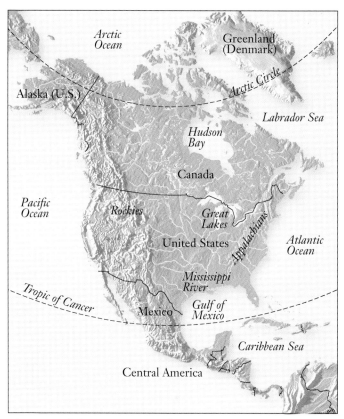

A map of the continent of North America, showing its major features. The Cordillera mountain system can be seen in the western area of the continent.

The geology

North America was the first continent to reach roughly its present shape and size—about 600 million years ago. It includes some of the oldest rocks in the world. The continent gradually grew around a vast area of ancient rocks called the Canadian Shield, which was formed nearly two billion years ago. Later, the Appalachian Mountains were forced up in the southeast. After this, a higher series of mountain ranges was formed in the west. Today called the North American Cordillera, these mountains stretch southeast from Alaska down through the Rocky Mountains into Central America.

In Arizona, the Grand Canyon reveals an almost complete geological history of the North American continent. The Colorado Plateau (pla-TOH; a flat area of raised land) gradually rose over a period between 10 and 5 million years ago. Since then, the steady flow of the Colorado River has gradually cut down through the layers of rock, leaving them exposed. At its deepest point, the canyon is 6,135 feet (1,870 m) deep, and the rocks at the bottom are more than one billion years old.

Like other continents, North America lies on one of the huge pieces called plates that make up Earth's crust. The North American Plate is slowly growing and moving westward. In the west, it meets the Pacific Plate beneath the Pacific Ocean, and in the east, the Eurasian Plate in the middle of the Atlantic Ocean.

Along the western coast of North America there are many earthquakes. These are caused by the movement of the North American Plate against the Pacific Plate. In the south, the San

The southwest of the United States is dry. Large areas of the landscape are desert in this area.

Andreas Fault line runs through the state of California from southeast to northwest, and out under the Pacific Ocean.

The movement of the North American and Pacific Plates sets up stresses in the rocks along the edges of the San Andreas Fault, and, from time to time, the stresses show on the surface as earthquakes. Most of these earthquakes are small and do little damage, but a major earthquake causes loss of life and destruction of property. Major earthquakes struck California in 1857 and 1906. Where the North American Plate meets the Eurasian Plate below the Atlantic Ocean, submarine volcanoes are forming new seafloor along the Mid-Atlantic Ridge, at a rate of about ¼ inch (6.4 mm) per year.

The land

The natural regions of North America can be divided roughly into six types that divide the continent from east to west.

Climate Regions

From north to south, the four major climate regions of North America are arctic, cool temperate, warm temperate, and tropical humid. The arctic region includes northern Canada, Alaska, and Greenland; it has only 60 frost-free days each year.

From east to west, the cool temperate region extends from Newfoundland to southern Alaska and south to Pennsylvania. A large area of forest stretches east to west across Canada in this region—the deciduous trees turning red and gold in fall (see picture). Winter is long and harsh, beginning at the end of October and lasting until early May. North of the Great Lakes, temperatures can drop to –80°F (–62°C) in February. Starting in spring, the frost-free growing season lasts 90 to 120 days. Moist air moving north from the warm temperate region brings rain in summer.

The warm temperate region lies in the southeast. It includes the Mississippi River basin and the Gulf Coast. Winters are mild with average February temperatures between 40° and 54°F (4.5° and 12°C). Rainfall is plentiful, and thunderstorms bring downpours in summer. In the deep southeast, hurricanes can be a hazard.

West of the warm temperate zone, into the Great Plains and Rocky Mountains, there is a mix of different climates—highland, steppe (dry, treeless area), and desert. Beyond the Rockies, in Oregon and California, summers are dry and crops have to be irrigated—supplied with water from the mountains.

Central America has a tropical climate, the temperature rarely dipping below 64°F (18°C). Winds blow onshore from the Atlantic Ocean, bringing plenty of rain. Hurricanes are a constant threat and cause great damage to crops and buildings, as well as loss of life.

On the east and south coasts, from Long Island, New York, to southern Mexico, the land is made up of plains and low hills. This area, together with the Caribbean, is threatened by hurricanes. The tropical islands of the Caribbean are the tips of a long mountain range rising from the floor of the ocean. They have narrow coastal lowlands and a mountainous interior.

The Appalachian Mountains run from Newfoundland, in northeast Canada, almost to the Gulf of Mexico in southeast United States. This very old range is lower than the Cordillera and has been worn down by the passage of time. The region contains major coal deposits.

North of the Great Lakes, the Canadian Shield is a vast region, with the smallest population in the continent. It is mountainous in the east, but low-lying inland and scoured by ice age glaciers. It is covered with small lakes and bogs that freeze over in winter.

Between the Appalachians in the east and the Rockies in the west, vast plains stretch from Canada nearly to the Gulf of Mexico. Most of the rivers flow southward to join the huge Mississippi River system. The western part, called the Great Plains, is drier than the east.

The western edge of the continent is made up of long mountain ranges, with high plateaus and narrow valleys. The Rocky Mountains, part of the Cordillera, run down the Canadian coast until, near the border with the United States, the Cordillera divides in two. The Cascades and the Sierra Nevada continue close to the coast, with the Rockies running farther inland and the Great Nevada Basin between the two.

CHECK THESE OUT!
✔AFRICA ✔AUSTRALASIA ✔CLIMATE ✔CONTINENT ✔EURASIA ✔PLATE TECTONICS ✔POLAR REGION ✔TEMPERATE REGION ✔TROPICAL REGION

Nova

While most stars shine steadily for millions of years, and some have steady and predictable changes in their brightness, a few are much harder to forecast. Novas are stars that occasionally show a sudden burst of brightness, often increasing their light many times over before fading away again. The word *nova* comes from the Latin for "new," because ancient astronomers thought that they were new stars appearing in the sky.

How a nova happens

As astronomers learned more, they realized that nova stars were there all the time (even if they could only be seen faintly through a telescope) and simply had occasional outbursts. Hundreds of novas are now cataloged, and astronomers think they have a good idea of how they work.

A nova happens in a binary star system, where two stars revolve around each other in a few days or less. They can only happen in systems where one star is much larger than the other and rushes through its life cycle while the other star is aging more slowly. As the massive star reaches the end of its life, it swells to turn into a red giant, puffing off its outer layers of gas, before collapsing into a tiny dense star called a white dwarf. The white dwarf has about the same mass as the Sun, but it is compressed into a ball only about the size of Earth.

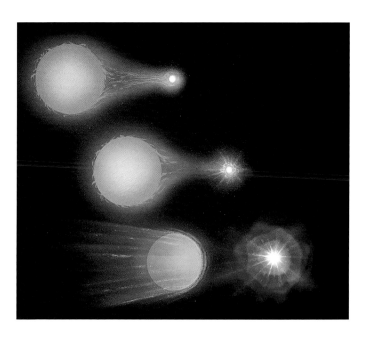

The atmosphere of a star is pulled toward a white dwarf star and causes an increase in brightness.

As thousands of years pass, the stars spiral closer to one another. In some cases, the white dwarf can get so close to the lighter star that it starts to pull the atmosphere away from that star. In most cases, however, this only happens once the lighter star itself reaches the end of its life and swells into a red giant. Eventually, the red giant gets large enough for its atmosphere to be pulled away by the white dwarf.

Hydrogen taken from the larger star spirals down onto the white dwarf itself, giving it a new atmosphere. The white dwarf's strong gravity starts to squeeze the hydrogen until eventually it begins to fuse as in normal stars, releasing large amounts of energy in the process. This nuclear fusion then makes the white dwarf's stolen atmosphere explode, causing a nova. Over a few hours, the star gets brighter before fading away over several days.

Supernovas

A supernova is a much bigger explosion than a nova. It completely destroys the exploding star and can outshine the light of an entire galaxy for

HIGHLIGHTS

- A nova is a star that brightens suddenly, appearing as a "new" star in the sky.

- Type I supernovas happen if the white dwarf in a nova system gains so much extra mass from the other stars that it collapses in on itself suddenly to form a tiny neutron star.

- Type II supernovas are supergiant stars exploding as they run out of fuel and collapsing violently.

several weeks. Supernovas only happen in Earth's galaxy roughly once in a century. Astronomers rarely get the chance to study supernovas with modern scientific equipment.

Supernovas can happen in one of two ways. Astronomers can tell which type it is from the supernova's light curve (the rate at which it brightens and fades away). Type I supernovas are closely related to novas. They happen in a star system where the white dwarf has a mass that is at the upper limit for this type of star. If their masses were any larger, the stars would not be able to support themselves, and they would collapse completely.

As hydrogen atmosphere pulled over from the nearby star builds up in the atmosphere of the white dwarf, it can tip the smaller star over the mass limit (roughly 1.4 times the mass of the Sun). The atoms inside the white dwarf suddenly collapse as the electrically charged protons and electrons inside them combine to form neutrons. Because the neutrons take up much less space, the white dwarf dwindles from the size of Earth to a neutron star the size of New York City. The whole process releases vast amounts of energy, and the star turns into a bright supernova.

Type II supernovas are caused by a different process. They happen at the end of a massive supergiant star's life cycle. As the star gets older, it generates energy by burning heavier and heavier elements in its core. Hydrogen burns to produce helium, helium generates carbon, and so on. The pressure of radiation pushing out through the star supports its outer layers, allowing it to swell to an enormous size. The cycle cannot go on forever. It ends with iron, which is too stable to burn. The star's power source suddenly disappears, the radiation holding up the outer layers is cut off, and the whole star collapses in seconds, releasing an enormous burst of energy, turning into a supernova as it forms a neutron star or an even denser black hole.

The brightness of a supernova depends on the mass of the original star. A Type II supernova can have a wide range of mass, so its brightness cannot be predicted. A Type I supernova always has the same mass, releases the same amount of energy, and therefore reaches the same brightness. Because astronomers can tell the two

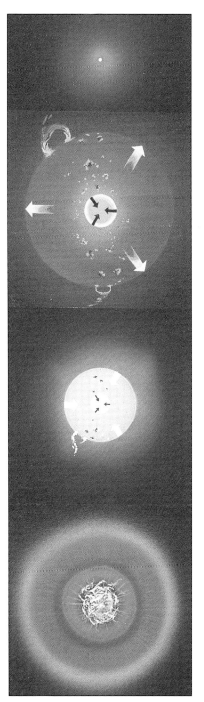

Phase 1: *A large, hot blue star burns hydrogen.*

Phase 2: *The star becomes a red giant, fusing helium atoms around the core.*

Phase 3: *Heavier and heavier atoms fuse. The star begins to shrink as its fuel runs out.*

Phase 4: *The density of the star becomes so great its protons and electrons are squeezed together to make neutrons, and the star explodes.*

The diagram above shows how a very large star can turn into a Type II supernova, which then explodes.

types apart, they can often identify Type I supernovas in distant galaxies and use them to work out the distance to the galaxies in which the supernovas appear.

CHECK THESE OUT!
✔BINARY STAR ✔RED GIANT ✔STAR ✔WHITE DWARF

Nuclear Physics

The branch of physics that studies the atomic nucleus and the particles inside it

The very heart of the atom—the tiny nucleus inside it—is investigated in nuclear physics. If an atom were the size of a football stadium, the nucleus would be the size of a pea at its center. In the century since it was discovered, scientists have learned a lot about the nucleus and released the power of nuclear energy.

Discovery of the nucleus

Chemists and physicists in the late 18th and 19th centuries had built up theories that explained the nature of matter based on tiny building blocks called atoms, even though they could not see them. Each element found in nature had its own unique atoms, and these atoms reacted together to form molecules and compounds. In 1897, British physicist J. J. Thomson (1856–1940) discovered that electrons (tiny particles with a negative charge) were part of each atom. He and others created the "plum pudding" model of the atom, which had electrons as plums scattered throughout a ball-shaped atom.

In 1911, one of Thomson's former students, Ernest Rutherford (1871–1937), decided to test this model. He used alpha particles (positively charged particles emitted by radioactive materials) to see whether the positive charge in

A particle accelerator in Switzerland. The ring shows where a tunnel runs underground. Scientists fire atomic nuclei around the tunnel to learn more about how protons and neutrons behave.

atoms was spread out as evenly as the plum pudding theory suggested. Rutherford fired a beam of these particles at thin gold foil. If the charge was distributed evenly, then they would pass straight through it. To his surprise, though, he found that while many particles went straight through, some were strongly deflected onto different paths, and a few bounced straight back out of the foil. The only explanation was that most of the atom's mass and positive charge was concentrated at the center of each atom, with the electrons placed around this. Rutherford named his discovery the nucleus.

Rutherford took his theory further, suggesting that the nucleus was similar to an alpha particle and was made up of smaller positively charged particles called protons. He discovered that the nucleus of hydrogen, the simplest element, contained one proton, and suggested that heavier elements contained more protons.

There was one problem with this model. Alpha particles turned out to be the nuclei (NOO-klee-EYE) of helium, the second lightest element, and scientists soon confirmed they

Nuclear Spin

LOOK CLOSER

Protons and neutrons are constantly moving around inside the nucleus, a bit like the way the electrons orbit around it. As a result, the nucleus can be said to rotate or spin. This nuclear spin can be used to identify different atoms and distinguish between isotopes with different numbers of neutrons.

Atomic nuclei can only spin with set amounts of energy. Scientists have discovered that spinning nuclei fall into a number of different energy levels when they are placed in a magnetic field. When atoms absorb radio waves, they can switch rapidly between these energy states in a process called nuclear magnetic resonance (NMR).

One widespread application of NMR is magnetic resonance imaging (MRI), which is used in hospitals. This technique uses magnetic fields and radio waves to record the magnetic resonance of spinning nuclei in cross sections inside the human body. The technique is very sensitive. Doctors use MRI to look at diseased areas of the body without having to operate on their patients.

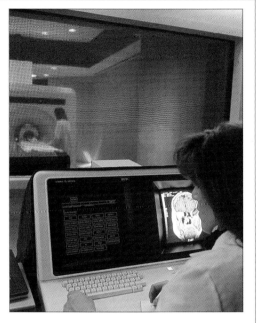

A doctor can look at a patient's brain using an MRI machine.

contained two protons. But if this was the case, why did helium weigh four times as much as hydrogen? Rutherford and other physicists soon realized that the nucleus must contain other particles, with the same mass as a proton, but no electric charge. This particle was named the neutron, which was eventually detected in 1932.

The nuclear model

Today, it is known that neutrons and protons (collectively called nucleons) are in turn made of even smaller particles. However, it is possible to understand atoms without referring to these particles. The number of protons in an element is absolutely unique—it defines which element an atom is. Because the overall charge on an atom has to be neutral, the positive protons in an atom are balanced by an equal number of negative electrons. However, positive protons would be expected to repel each other, causing the nucleus to fly apart. The reason it does not do this is that, inside the nucleus, the electromagnetic forces that make similar charges repel are overwhelmed by much stronger nuclear forces, which act only over very short distances.

Neutrons, meanwhile, add mass to the nucleus. Nuclei of the same element can contain different numbers of neutrons, and the resulting atoms, with a range of different masses, are called isotopes. In nature, the range of isotopes of any element is limited, because the numbers of protons and neutrons are roughly balanced.

When the number of protons and neutrons in an atom gets out of balance, the nucleus becomes unstable. Unstable atoms lose particles of one sort or another to become stable again. This is called radioactivity.

In 1934, Italian scientist Enrico Fermi (1901–1954) began to deliberately bombard atoms with alpha particles and then with individual neutrons. Fermi discovered the rules that governs the transformation of one element into another by nuclear reactions. He was eventually able to manufacture atoms of new elements, and even isotopes that were highly unstable and therefore radioactive. Fermi's experiments paved the way for the nuclear age, with the problems and promise of nuclear power and the threat of nuclear weapons.

CHECK THESE OUT!
✔ATOM ✔ELECTRON ✔FISSION ✔FUSION ✔ISOTOPE
✔NEUTRON ✔PARTICLE PHYSICS ✔PHYSICS ✔PROTON
✔RADIOACTIVITY ✔SUBATOMIC STRUCTURE

Ocean

Body of saltwater covering Earth's surface

The oceans cover more than 70 percent of Earth's surface and hold 97 percent of the planet's water. They are home to a wide variety of living organisms, and they provide humans with food, energy, and valuable minerals. The oceans are also the source of the fresh rainwater that makes life on land possible.

Earth's four oceans, in descending order of size are: the Pacific Ocean, the Atlantic Ocean, the Indian Ocean, and the Arctic Ocean. The Antarctic Ocean, or Southern Ocean, consists of the southern parts of the Pacific, Indian, and Atlantic Oceans. Seas are smaller bodies of saltwater and are usually located close to, around, or between large landmasses. Apart from inland seas, most seas are part of a larger ocean.

HIGHLIGHTS

- ◆ The oceans cover more than 70 percent of Earth's surface and hold 97 percent of its water.

- ◆ The four oceans are the Pacific, Atlantic, Indian, and Arctic Oceans.

- ◆ The size and shape of the oceans are changing all the time.

- ◆ The ocean floor is a landscape of mountains, valleys, cliffs, trenches, and vast plains.

- ◆ Ocean waves are continually reshaping the coasts, by wearing the land away or building it up.

Birth of the oceans

When Earth first formed, it was a hot, barren planet, but its atmosphere contained water vapor. As Earth cooled, the water vapor in its atmosphere condensed (turned from a gas into a liquid) and fell as rain. Over time, the rainwater collected in basins to become the oceans. Today, this same water continues to cycle from Earth's surface up into the atmosphere and back down again as rain, snow, or hail, through the processes of evaporation (turning from a liquid into a gas) and condensation (turning gas into a liquid).

Since the oceans first formed, their shape and size have continually changed. This is because the huge plates that make up Earth's crust are not still, but move very slowly over the liquid layers of rock beneath. The plates carry continent or ocean floor, or both. As the plates move, they change the shape of the oceans.

The ocean floor

Beneath the shallow seas that surround the continents is the area called the continental margin. Moving out from the land, this is made up of the continental shelf, the continental slope, and the continental rise. The continental shelf slopes gently down toward the continental slope, which plunges steeply down toward the continental rise. The rise is a thick layer of sediment (particles of rock and other materials deposited from the water) between the continental slope and the deeper ocean floor.

Beyond the continental margin, the ocean floor is a dramatic underwater landscape—a vast plain broken by mountains, valleys, cliffs, and

LOOK CLOSER

Features of the Ocean Depths

The ocean depths (the abyss) contain many dramatic natural features. Steep-sided trenches and canyons descend to great depths. In the Pacific Ocean, the Mariana Trench drops to a depth of 36,198 feet (11,033 m)—the deepest point on Earth. Ocean trenches are mostly formed by movements of Earth's tectonic plates. Where two ocean plates collide with each other, one is pushed beneath the other, forming a deep trench. This process is called subduction.

Earth's deep waters contain ridges and high mountains. For example, seamounts more than 3,300 feet (1,000 m) high are very common in the Pacific. In some places, they break the surface to form islands. In the tropics, these islands are often surrounded by coral reefs. Abyssal hills, smaller and more numerous than seamounts, do not reach the ocean surface.

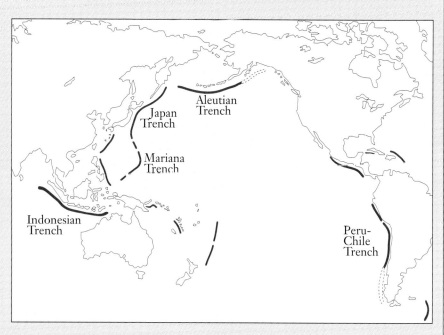

A map showing major trenches (as red lines) in the floor of the Pacific Ocean. Dotted lines show trenches that are filled with sediment.

steep-sided underwater gorges called trenches. Nearly one-third of the ocean floor is taken up by ridges (chains) of underwater mountains, with most of their peaks lying 1 to 2 miles (1.6 to 3.2 km) below the ocean surface.

Ocean currents

The waters of the oceans are always in motion. Their movement is driven by winds blowing across their surface and by collisions between masses of warm and cold water. In their turn, the ocean currents influence the world's climates. Ocean currents are like huge rivers, with paths and patterns of flow that can be predicted. For example, warm water from the tropics surges northward and southward to moderate the winters in higher latitudes.

Surface currents are driven by the wind. The strongest currents flow along the western borders of the oceans. In turn, surface currents affect the circulation of the deep waters, forcing the waters to upwell (rise) and downwell (sink).

Shaping the coasts

The coasts that edge the continents are constantly being reshaped by two main forces—erosion (wearing away) and deposition (laying down of loose, rocky sediment, such as sand and gravel). Ocean waves are a major force of erosion; they hurl sand and pebbles that act like sandpaper, scraping away at the coast. The crashing waves eat away at the land to create cliffs and other features, depending on the hardness of the coastal rocks.

Where waves erode (wear away) the coast, beaches are usually made up of pieces of the eroded rock. Other beaches are made up of sediment deposited by rivers or ocean currents, or from materials such as coral.

CHECK THESE OUT!
✔CONTINENTAL SHELF ✔EROSION ✔GULF STREAM
✔HYDROLOGY ✔MEDITERRANEAN SEA
✔OCEAN CURRENT ✔OCEANOGRAPHY
✔PLATE TECTONICS ✔WATER ✔WAVES

Ocean Current

**The patterns of moving water that flow
up, down, or across the ocean**

The waters of the ocean are always moving, as currents flow like rivers through the water, waves crash on beaches, and tides rise and fall. Ocean currents circulate the waters of the ocean in much the same way as winds move air.

Ocean currents are streams that move water around the globe as well as from the surface of the ocean down to the seafloor and vice versa. Ocean currents can transport vast quantities of water. One of the best known currents is the Gulf Stream, which flows along the east coast of the United States, carrying warm, tropical water via Canada to northwest Europe.

Surface currents

Surface currents flow within several hundred feet of the ocean surface. They flow mainly horizontally (along, not up or down) and are

HIGHLIGHTS

◆ Ocean currents circulate the waters of the oceans.

◆ The main types of ocean currents are surface currents, density currents, and longshore currents. Surface currents are driven by winds.

◆ Density currents are caused by differences in water temperature and salt content.

◆ Ocean currents have important effects on the weather and the organisms that live in the seas.

Earth's main ocean currents. The blue arrows show colder currents, the red ones show warmer currents.

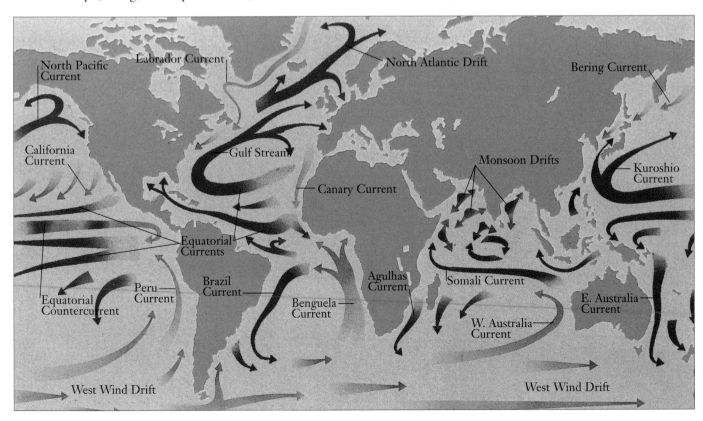

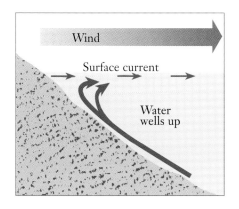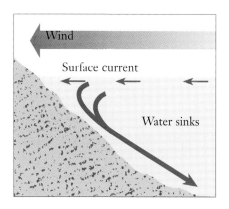

This diagram shows how wind can cause vertical currents along a coastline. When wind blows out to sea, it sets up a surface current that draws water up from the seafloor. When wind blows inland, it makes a downwelling current.

driven by winds blowing across the ocean surface and moving the water forward. Surface currents follow set paths, and these paths follow the pattern of the prevailing winds (the winds that tend to blow most of the year), which might vary slightly with different seasons.

Surface currents are generally strongest in regions of strong, steady winds, such as those that blow toward the equator. Like winds, surface currents are deflected (bent) by Earth's spin. Earth is spinning toward the east, and, in the Northern Hemisphere, this deflects winds to the west of their path. The winds push the surface water 45 degrees to the right of the direction in which the winds are blowing, which makes the water flow in a clockwise, circular pattern. In the Southern Hemisphere, winds are also deflected to the west of their path. The surface water is pushed a farther 45 degrees to the left, creating a counterclockwise flow. These clockwise and counterclockwise circular flows are called gyres (JY-uhrz). All the waters of Earth's oceans spin around in separate gyres in the Northern and Southern Hemisphere.

The power of a surface current depends on the strength of the winds and the width of the current (narrow currents flow faster than wide ones). In addition, Earth's eastward rotation makes surface currents stronger along the western coastlines. Surface currents can flow as quickly as 3 feet (1 m) per second.

Density currents

Density currents flow vertically (up and down) from the surface of the ocean to the bottom, and back again. They are usually slower than surface currents, and it can take as long as 1,000 years for water to move from the depths to the surface.

Density currents are driven by differences in water density, which are caused by variations in the temperature and saltiness of seawater. Colder water is more dense than warmer water, and salt makes water even denser. Temperature differences occur because the Sun heats the ocean to higher temperatures in tropical regions than in polar regions. Changes in saltiness are caused by fresh water entering from rivers and leaving the ocean by evaporation.

Vertical currents start when dense water sinks and less dense water rises to take its place. Areas where water sinks are called downwelling zones. Upwelling zones are places where water rises.

Downwelling and upwelling also happen where surface ocean currents converge (meet) or diverge (flow apart). Where two or more currents converge, water sinks because it has nowhere else to go. Where currents diverge, water from below wells up to fill the space.

Density currents play a vital role in the life of the ocean. Downwelling carries oxygen-rich surface waters to deep-sea creatures. Upwelling brings nutrients up to the surface, where they feed the living organisms that live there.

Longshore currents

As their name suggests, longshore currents move along coasts. They happen when waves approach the shore at an angle and then change direction as they enter shallow water. The strength and speed of longshore currents are affected by the size of the waves and the amount they turn.

CHECK THESE OUT!
✔DENSITY ✔EARTHQUAKE ✔EL NIÑO AND LA NIÑA
✔OCEAN ✔OCEANOGRAPHY ✔POLAR REGION
✔TIDE ✔TROPICAL REGION ✔WAVES ✔WIND

Oceania

The area made up of Australia, New Zealand, and more than 10,000 islands scattered across the Pacific Ocean is called Oceania (OH-shee-AN-ee-uh). Most geographers agree that Oceania includes four main areas: Australasia (Australia and New Guinea), Melanesia, Micronesia, and Polynesia. However, the borders of Oceania are defined by the culture and history of its peoples, as well as by geography, and this has led to some disagreement. Some experts claim that the Aleutian Islands in the north and the islands of Japan off eastern Asia are part of Oceania because they are Pacific island groups. Indonesia, the Philippines, and Taiwan have also been considered part of Oceania.

History of Oceania

Scientists have discovered that Oceania was one of the last areas of Earth to be settled by humans. Experts believe that Australasia was first inhabited about 40,000 years ago, and many of the smaller islands may have been inhabited for only a few thousand years or less. Oceania's first inhabitants probably came from southeast Asia.

In 1521, an expedition of sailors led by Portuguese explorer Ferdinand Magellan (around 1480–1521) were the first Europeans to make contact with the people of Oceania. During the next 250 years, European explorers encountered the inhabitants of nearly all the islands in the region. By the early 1800s, a few European sailors and missionaries had settled to live on the larger islands.

During the 1900s, air transport increased links between the islands of Oceania and the Americas and Europe. Oceania gradually came to play a greater part in world economy and politics. During World War II (1939–1945), bitter fighting throughout Oceania focused world attention on the region for the first time in modern history.

Australasia

In the past, the area now considered to be Oceania was called Australasia, which means "south of Asia." Today, this term is generally used to mean Australia and New Guinea only, but it sometimes includes New Zealand.

Together, Australia and New Guinea total 3.5 million square miles (8.5 million sq km). Most of Australia is a gently rolling plateau (pla-TOH; high flat area). The highest mountain is Mount Kosciusko, at 7,316 feet (2,230 m). In contrast, New Guinea is a mountainous land covered by dense tropical rain forests. Some of its highest mountains are snow-capped throughout the year, despite the fact that the island lies just south of the equator (ih-KWAY-tuhr; an imaginary circle around Earth at equal distances from the poles).

The original inhabitants of Australasia probably arrived from south Asia about 40,000 years ago. Today, only about 160,000 of Australia's 15 million people are descended from those first inhabitants.

Melanesia

Melanesia is a 2,000-mile (3,220-km) chain of islands including the New Hebrides, Solomon, New Caledonia, and Fiji Islands. The land totals

---

HIGHLIGHTS

♦ Oceania includes four distinct regions: Australasia, Melanesia, Micronesia, and Polynesia.

♦ Oceania was one of the last areas on Earth to be inhabited by humans.

♦ The area called Oceania today was once called Australasia (meaning "south of Asia").

♦ Most of the thousands of tiny islands in Oceania were formed by volcanic eruptions.

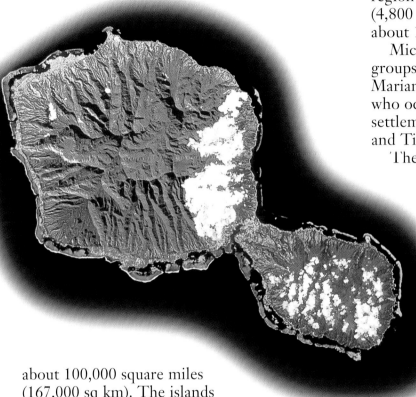

A satellite image of Moorea (left) and Tahiti (below), both part of the Society Islands chain in Polynesia.

about 100,000 square miles (167,000 sq km). The islands were formed mainly by a group of undersea volcanoes that erupted long ago. Eventually, the volcanoes erupted so much lava that the rocky masses broke the sea surface to form islands. Over time, coral reefs have formed in the coastal waters fringing the islands.

While the islands of Melanesia are mainly low-lying, some of the deepest ocean trenches in the world lie just offshore. The Mariana Trench, east of the Mariana Islands in the western Pacific, plunges to 36,198 feet (11,033 m) below sea level—the deepest point on Earth.

The origins of the people of Melanesia are uncertain, but many experts believe humans first reached the area from Asia, across a bridge of land that was later submerged under rising seas. Today, Melanesia has about 1.25 million people. Many of the region's islands are uninhabited.

Micronesia

About 2,250 islands make up Micronesia. They are scattered over an area roughly the size of the United States. However, their total land area adds up to only 1,100 square miles (2,875 sq km), which is about one-fifth the size of Connecticut. Micronesia lies northeast of New Guinea. The region stretches westward for about 3,000 miles (4,800 km) and northward from the equator for about 1,000 miles (1,600 km).

Micronesia is made up of several island groups, including the Marshalls, Carolines, and Marianas. The region has 260,000 inhabitants, who occupy 110 of the islands. The largest settlements are on the islands of Guam, Saipan, and Tinian in the Marianas.

The islands of Micronesia are mostly the remains of an ancient volcanic ridge that runs toward Japan. Nearly all the islands are now surrounded by coral reefs.

The physical features and culture of the Micronesians are different from those of the peoples of Polynesia and Melanesia. Experts believe that the original inhabitants of Micronesia came from the Philippines or from Malaysia.

Polynesia

Polynesia is the largest region of Oceania, covering 15 million square miles (39 million sq km) of water. However, the total land area is only 113,500 square miles (294,000 sq km). Polynesia stretches south from the Hawaiian Islands to New Zealand and east to Easter Island. Most of the islands began life as undersea volcanoes. Volcanic activity continues in New Zealand and the Hawaiian Islands.

The origins of the people of Polynesia are uncertain. Some experts believe that they came from Asia or Melanesia. Others note the cultural connections and the similarity of blood types between Polynesians and Native Americans. They believe that the two groups of people had close links in the distant past. Today, more than four million people live in Polynesia. Around 90 percent of them live in either New Zealand or the Hawaiian Islands.

CHECK THESE OUT!

✔AUSTRALASIA ✔ISLAND ✔PACIFIC OCEAN ✔VOLCANO

Oceanography

The study of the physics, chemistry, geology, and biology of oceans and ocean floors

The Hubble Space Telescope is used to explore distant galaxies, and the Galileo mission to Jupiter continues to probe deeper into Earth's Solar System. However, the oceans that cover 71 percent of Earth remain almost completely unexplored.

Oceanography is the study of the oceans. Greek philosopher Aristotle (384–322 B.C.E.) is often said to be the founder of this science. He classified dozens of sea inhabitants, including sponges, clams, and oysters. He also

HIGHLIGHTS

◆ Greek philosopher Aristotle is credited as the founder of oceanography, but modern scientific studies began only in the 19th century.

◆ Diving equipment, scuba gear, and very strong submersibles have made exploration of the ocean's depths easier.

◆ Present-day oceanography is costly, and many nations have joined together in major programs.

◆ A knowledge of ocean circulation is important in weather forecasting.

distinguished warm-blooded, air-breathing animals from cold-blooded fish that breathe through gills, and even described the sound waves dolphins use as they move about.

Nobody knows who supplied Aristotle with marine specimens. It may have been his student, Alexander the Great (356–323 B.C.E.). In 333 B.C.E., Alexander ventured into the sea in a primitive diving bell. This held an air bubble in its dome so that the diver could breathe.

The first scientific study of the ocean's movements was made by U.S. Navy officer Matthew Fontaine Maury (1806–1873). He was superintendent of the Navy's Depot of Charts and Instruments, and he collected information from navigators who had traveled all over the world. He used that information to describe currents and the wind's effect on the oceans.

Many scientists believed that animals could not survive the great pressure of the ocean's

A 16th-century painting of Alexander the Great using a diving bell in Lebanon 2,000 years earlier.

depths. Then, in the 1860s, U.S. geologists brought up worms and mollusks from a depth of 4,000 feet (1,200 m). Years later, when a broken telegraph cable was raised from a depth of 7,000 feet (2,100 m), it was found to be encrusted with types of marine life.

The most thorough 19th-century study of the ocean began in 1872. The British vessel HMS *Challenger* cruised the globe for more than three years. The scientists aboard took water samples, temperature readings, depth recordings, and plant and animal specimens. They collected more than 4,700 new species and proved that some were adapted to life at the greatest depths. They also discovered the Mariana Trench in the Pacific Ocean, which is the ocean's deepest place. They recorded a depth of 26,850 feet (8,184 m), while scientists now believe that the trench's deepest point reaches 36,198 feet (11,033 m).

The *Challenger*'s scientists produced a report consisting of 50 volumes, nearly 30,000 pages long. This document gave birth to the modern science of oceanography.

Early diving bells held only a limited amount of breathable air, so the time that could be spent underwater was short. During the early 1800s, diving suits with supplies of pumped air were invented. These were heavy and clumsy, and the length of the air hose meant that divers could not move very far. In the early 1930s, U.S. naturalist Charles William Beebe (1879–1961) developed a spherical diving vessel called a bathysphere (BA-thihs-fihr). In 1934, Beebe made a record-breaking dive in the bathysphere to a depth of 3,028 feet (923 m).

In the 1940s, French oceanographer Jacques Cousteau (1910–1997) invented the *self–contained underwater breathing apparatus* (scuba). With air tanks mounted on their backs, divers were able to swim freely. Cousteau received generous funding from the French government and the National Geographic Society. He was able to bring the wonders of the deep into the homes of millions, through his many movies, television shows, and books.

Scuba divers were still limited by water pressure and could go no lower than around 165 feet (50 m). To explore the ocean's greater depths, engineers in the 1960s began to build

EVERYDAY SCIENCE

Influencing the Weather

Because they cover 71 percent of Earth's surface, the oceans have a major influence on the weather. As people learn more about the way in which ocean water circulates, they are able to make better predictions of weather changes.

It is only since the launch of weather satellites that oceanographers have been able to study patterns of ocean circulation. Satellites use cameras and other detectors to collect information such as sea surface temperature, surface wind, and ocean color. This information is sent by radio to supercomputers on Earth and translated into color-coded maps. The maps show warm and cold ocean currents, wind speeds, and swirling ocean eddies.

To predict the coming weather, researchers use computer models. Data on past ocean and weather patterns is fed into the computer. The computer can then be used to model changing weather conditions as it receives new information. The pattern of moving clouds shown in weather forecasts on television is an example of computer modeling.

Several nations have joined together on long-term projects to study the connection between ocean circulation and weather. One of the main areas of research are the El Niño and La Niña events, which affect weather patterns all over Earth.

A computer model of ocean temperature. The red parts are warm areas, the blue are cold.

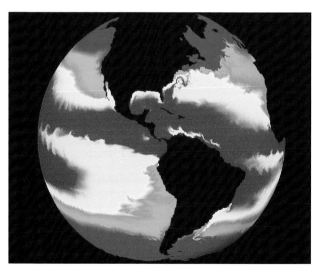

small, one- and two-person submarines (submersibles). These research vessels have plunged as deep as 11,000 feet (3,353 m). Using cameras and movable arms, the operators can record and sample deep-sea life.

Modern oceanography

Research equipment is now very complex, and the cost of ocean studies is high. In the second half of the 20th century, oceanography became concentrated at only a few large institutions. These include the Woods Hole Oceanographic Institution in Massachusetts, Scripps Institution of Oceanography in California, and Plymouth Marine Laboratory in England.

Long-term ocean research projects often use scientists from several nations. The Ocean Drilling Project involves 20 countries. Researchers aboard the program's drilling ships study the geological history of the ocean floor and the structure of the tectonic plate

Oceanographers prepare a probe that has been designed to follow deep ocean currents.

boundaries. The Ridge Inter-Disciplinary Global Experiments investigate the movements of these plates to help predict volcanic eruptions, earthquakes, and tsunamis (soo-NAH-meez; tidal waves).

At least two major international programs are looking into the effect of waste gases, from the burning of fossil fuels and other pollutants, on marine life. The Joint Global Flux Study is concerned with the impact of global warming. Scientists taking part measure changes in tiny floating organisms, such as algae (AL-jee) that make up plankton (tiny animal and plantlike organisms). Blooms are sudden increases in the population of plankton caused by favorable conditions such as warm sea temperatures, a large food supply, and the water's mineral content. The Global Ocean Ecosystem Dynamics Program looks at the reasons for variations in the population of marine animals.

A submersible is prepared to take three crew members to a depth of 4,000 feet (1,220 m).

EXPLORERS

Sylvia A. Earle (born 1935) is one of the world's leading deep-sea explorers. In 1952, as a 17-year-old student, she made her first scuba dive. She earned her master's degree at the age of 20, and, as a doctoral student at Duke University, she concentrated on the study of ocean-living algae.

Sylvia Earle made history in 1970. She led a team of five women scientists who spent two weeks living in two huge tanks, 50 feet (15 m) deep in the ocean off the U.S. Virgin Islands. The project, named *Tektite II*, was designed to see how well people could perform in such conditions. The team adapted well, and spent most of their waking hours exploring their underwater surroundings. Earle cataloged 153 species of plants, 26 of which had never been seen previously in the Virgin Islands. In 1979, she plunged alone to a depth of 1,250 feet (375 m), making a dangerous, record-breaking dive in a suit of metal and plastic called a Jim suit. Earle spent three hours walking on the ocean bed collecting specimens with Jim's robotic arms, and she planted a U.S. flag.

In 1982, Earle and engineer Graham Hawkes founded Deep Ocean Engineering, a company to design and build submersibles. They created *Deep Rover*, a one-person submersible, and *Deep Flight*, a vessel designed to "fly" through the ocean at depths greater than 4,000 feet (1,200 m). In 1990, Earle became the first woman to be appointed chief scientist to the National Oceanic and Atmospheric Administration (NOAA).

Sylvia Earle shows a sea creature to a colleague inside the Tektite II *tank.*

Today, oceanographers are as likely to be sitting in front of a computer on land, as they are to be on a ship at sea. Much of their data now comes from satellites. The National Oceanic and Atmospheric Administration (NOAA) began launching these in the late 1960s. Instruments onboard can record such information as the amount of salts in the seawater; temperature changes; plankton distribution; and the flow of waves, ocean currents, and tides. Computers convert the data into charts and maps.

Apart from its importance to ocean studies, this data also has practical uses. Maps showing concentrations of plankton can help commercial fishers to increase their catch. Military submarines can use data on differences of water density to choose routes through the ocean on which they can travel without being detected.

Astronauts can see large-scale variations that are not visible on Earth's surface. For many years, oceanographers believed that whirlpools in the ocean water were rare. Then, in 1984, astronauts aboard the space shuttle were able to take photographs showing that, all over the world, there are spiral eddy currents. These can cover an area as large as the state of Rhode Island and last for many years.

Modern oceanography is divided into four main types, and scientists specialize in one of them. Physical oceanographers study the movements of the ocean. Chemists are concerned with the chemical composition of seawater, which contains at least 80 different elements. Ocean geologists study the formation of the ocean floor. Marine biologists are concerned with the living organisms that live in seawater.

Many marine scientists must still go to sea. Large research ships can accommodate dozens of scientists, who often spend months aboard. Marine biologists often make their most important observations underwater. Scuba gear and submersibles are among their most valuable equipment. Ocean geologists often travel aboard drilling ships, which can collect samples of the ocean bed at depths of several thousand feet.

CHECK THESE OUT!
✔EL NIÑO AND LA NIÑA ✔MARINE EXPLORATION
✔OCEAN ✔OCEAN CURRENT ✔TIDE ✔WAVES

Oil

The most important fossil fuel, used to power modern industry

During the 20th century, oil became probably the world's most important natural product. It is still essential for nearly all current forms of transport and has brought heat and electricity to the whole industrialized world. It is the raw material for many industries manufacturing such products as plastics, fertilizers, detergents, paints, solvents, and explosives.

The existence of petroleum (oil and natural gas) has been known since ancient times. In some places, oil came to the surface and evaporated, leaving tar (a black sticky substance). This was used in medicines and the preserving of bodies. In parts of the Caspian Sea in Central Asia, oil and gas bubbled up to the water surface.

By the end of the 18th century, Myanmar (Burma) had more than 500 wells producing around 40,000 tons (35,700 tonnes) of oil per year. Much of this provided lamp and stove oil. By the end of the 19th century, millions of tons were being extracted every year from around the Caspian Sea and Titusville, Pennsylvania.

However, it was not until the development of electric lighting and the invention of the internal combustion engine that oil became the world's

An oil rig in Texas pumping out petroleum from beneath the ground.

most sought-after fuel. The consequent search for oil deposits—and the huge profits to be made—gave it its nickname of "black gold."

The word *petroleum* means "rock oil," and this reflects its occurrence in the pores of sedimentary rocks. Like coal and natural gas, oil is a fossil fuel, formed from the decayed remains of ancient sea-dwelling organisms, such as shellfish and algae (AL-jee; plantlike organisms). These remains became covered by mud, and as the layer of mud increased in depth, the pressure above caused it to harden into rock. The algae and animal remains were forced deep below

HIGHLIGHTS

◆ Oil is formed from the decayed remains of ancient marine organisms.

◆ Oil forms in sedimentary rocks and seeps into the tiny holes in certain types of rocks.

◆ Crude oil is recovered and brought to the surface by drilling through reservoir rocks.

◆ Around half of all petroleum products are used as fuel for transportation.

ground, where a series of physical and chemical changes took place.

Oil is a mixture of hydrocarbons (compounds of hydrogen and carbon). Some of these come naturally from decayed remains, and others result from the effects of pressure and Earth's heat on the chemicals inside the dead organisms.

Searching for oil

Over time, petroleum has seeped from its source into tiny holes inside certain rocks, making a reservoir. Natural gas often bubbles to the surface of the oil, and the oil itself may float on

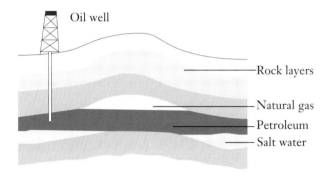

Oil well

Rock layers

Natural gas

Petroleum

Salt water

This diagram shows oil trapped underground.

salty groundwater. Some reservoirs are under pressure, and when a drill cuts through the overlying rock, the oil rushes upward. Other oil reservoirs have to be pumped out.

To avoid pointless drilling, geologists (scientists who study the structure and history of rocks) have to study the rock strata (STRAH-tuh; layers). Sometimes they set off explosives and watch the way in which the shock waves are reflected to locate any petroleum below. Many oil fields lie partially or wholly under the ocean. About one-third of all world oil comes from wells that have been drilled off shore.

Oil is made almost entirely of hydrocarbons. However, there are many different compounds, and they must be separated before they can be used. This process is called refining. Refining separates out fractions (families of chemicals) and removes impurities such as sulfur.

CHECK THESE OUT!
✔ELECTRICITY ✔HYDROCARBON ✔NATURAL GAS
✔PETROCHEMICAL ✔REFINING ✔SEDIMENTARY ROCK

LOOK CLOSER

Earth's Oil Reserves

Oil supplies are calculated in barrels. A barrel is around 35 gallons (132 liters). Earth's oil reserves are estimated at around one trillion barrels. More than half are located in just four countries: Saudi Arabia, Iraq, Kuwait, and Iran. Other members of OPEC (the Organization of Petroleum Exporting Countries), such as Venezuela, make up another 25 percent of the reserves.

In the United States, the Gulf Coast area of Texas is the most important oil-producing area. The discovery of Spindletop in 1901 was considered the greatest find of all time. Oil flowed at a rate of 80,000 barrels a day, and the discovery set off an immediate oil rush. However, within a year the flow from the entire Spindletop field was down to 5,000 barrels a day.

World consumption of oil is currently around 80 million barrels a day. Many people are worried about the inadequate supply of reserves, and the fact that so much oil is concentrated in the Middle East. Some countries have already begun to introduce energy-conservation programs, and to substitute natural gas for petroleum or develop new sources of energy, such as sugar or sunlight. However, the industrialized world still depends heavily on oil. Calculations have shown that Earth's oil supplies will begin to run out rapidly around the year 2030.

Beneath this pumping station in Texas, a reservoir of oil is stored in case of emergency.

Optics

The study of how light behaves and how it can be used in everyday life

Light is remarkable. It is responsible for vision, and it plays an increasingly important part in many modern technologies. Lasers, laptop computers, and fiber optics all rely on the properties of light.

The study of light and how it behaves is called optics. This is a very broad branch of physics and includes everything from what light is and how it travels, to the design and construction of very advanced optical devices such as the Hubble Space Telescope.

What is light?

Light is a type of electromagnetic radiation. Like X rays and infrared radiation, it travels at a speed of 186,282 miles per second (299,792 km/s). People can see objects because light from the Sun reflects off them into people's eyes. Inside

The skyline of New York City is reflected in water. Reflection is one of the ways light behaves.

HIGHLIGHTS

◆ Optics is the study of how light is created, produced, and transmitted from place to place.

◆ Light's double nature—part particle and part wave—explains many of its properties.

◆ Reflection, refraction, diffraction, and interference are the most important properties of light.

◆ The study of optics is important in many modern technologies, from designing telescopes to storing information on compact discs.

Reflection and Refraction

LOOK CLOSER

Reflection and refraction are two of the most important properties of light. Reflection is what happens when a beam of light hits something and bounces back again. Refraction is the way that light bends when it passes through something.

Reflection explains why some things appear colored. A tomato looks red because it reflects only red light. When ordinary white light strikes a tomato, the red part of that light is reflected into people's eyes, while the other colors of light are absorbed. A piece of paper looks white because it reflects all the light that falls onto it. In contrast, a piece of coal looks black because it absorbs all the light falling on it and reflects none of it.

When people look in a mirror, the image they see is formed according to the law of reflection, which states that light rays reflect from a smooth surface at the same angle at which they strike the surface. The light striking the viewer's eye seems to come from an image that is as far behind the mirror as the object is in front of it, and the image is flipped so that the left hand appears on the right side.

Refraction (the bending of light) happens because light travels at different speeds as it moves through different substances, such as air and water. It is why a straw in a glass of water appears bent (see picture).

eyes, a lens brings the light rays into focus on a sensitive light-detecting surface at the back of the eye called the retina (REH-tuh-nuh). The retina converts the light rays into electrical signals that the brain then turns into a picture.

Ordinary light is sometimes called white light, even though it is made up of a mixture of colored light. Light is usually thought of as a wave. Light of different colors travels at the same speed but has a different wavelength (the distance between one wave crest and the next) and frequency (the number of waves that arrive each second). When light passes from one medium to another, the wavelength changes but the frequency does not. Sometimes it is possible to see the colors in white light. If a CD (compact disc) is held toward the light and turned slightly, the spectrum of colors can be seen. This happens because the surface of the CD splits white light into its component colors and bends each one toward a person's eyes by a different amount.

The same thing happens when a beam of white light enters a prism (PRIH-zuhm; a triangular block of glass) and comes out of the other side as a spectrum (rainbowlike pattern of colored light).

Light behaves like a wave, but sometimes it can also behave like a stream of particles. These tiny light particles are called photons (FOH-tahnz). Each photon is a tiny packet of energy. Light is both a wave and a particle—this strange idea is important in the more advanced study of light. Scientists call it the wave–particle duality (being a wave and particle at the same time).

How light behaves

Light's wavelike nature is responsible for many of its properties. A beam of light will be reflected from objects that get in its way. Light is thought of as reflecting only off shiny surfaces. However, light of one wavelength or another will reflect off most things, and this is the reason that people can see different colors.

Light travels at different speeds in different substances. It travels more slowly through water or a block of glass than through air, and more slowly through air than in a vacuum. When a light strikes the surface of a new material, the difference in speeds causes it to refract (bend). This explains why a stick appears to be bent when it stands half in and half out of a pool of water. On a hot day, what looks like a body of water can be observed on the surface of a road. This pool is actually an image of the sky superimposed on the road. The image, called a mirage (muh-RAHZH), is caused by hot air refracting the light from the sky back up into people's eyes. Reflection is a type of refraction where light enters something like a piece of glass and bends so much that it comes back out.

Apart from reflection and refraction, there are two other important properties of light: interference and diffraction. Interference is what happens when two or more light waves meet one another. Diffraction explains how light waves spread out when they pass through a small opening or around a sharp edge.

Optics in action

Reflection and refraction are used in the simplest types of optical equipment, such as lenses and mirrors. A lens is a piece of glass or plastic that bends (refracts) light rays as they pass through it. A convex (converging) lens bends the rays inward so they come to focus at a point. The lenses in human eyes and in glasses for farsighted people work like this. A concave (diverging) lens bends light rays so that they spread apart. It is used to help nearsighted eyes.

Just as lenses rely on refraction, so mirrors use reflection. Most mirrors are made of a highly polished piece of metal attached to the back of a piece of glass. The idea is to reflect back as much of the light that hits the mirror as possible. Scientific instruments use even more carefully designed mirrors. Laboratory experiments with lasers usually require scientists to use delicate mirrors made by coating a surface with grains of silver. The mirror used in the Hubble Space Telescope took several years to construct and is undoubtedly one of the most precise scientific instruments ever produced.

Optics in the modern world

Optics is not just a science; it is also a technology: a science that can be used to solve everyday problems. The simple laws of optics that explain reflection and refraction are used to design many optical devices. Microscopes, telescopes, binoculars, and cameras are all made using lenses and mirrors that reflect or refract light. An ordinary camera, for example, has a lens to look through called a viewfinder and a second lens at the front that brings the picture into focus on the light-sensitive film inside. A more sophisticated type of camera called a single-lens reflex (SLR) uses a flip-up mirror so that the picture seen through the viewfinder is the same as the picture that appears on the film.

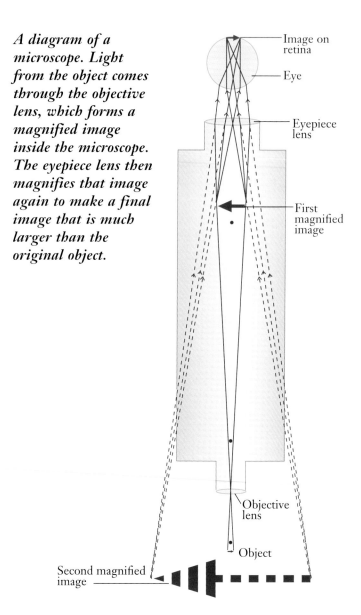

A diagram of a microscope. Light from the object comes through the objective lens, which forms a magnified image inside the microscope. The eyepiece lens then magnifies that image again to make a final image that is much larger than the original object.

Image on retina

Eye

Eyepiece lens

First magnified image

Objective lens

Object

Second magnified image

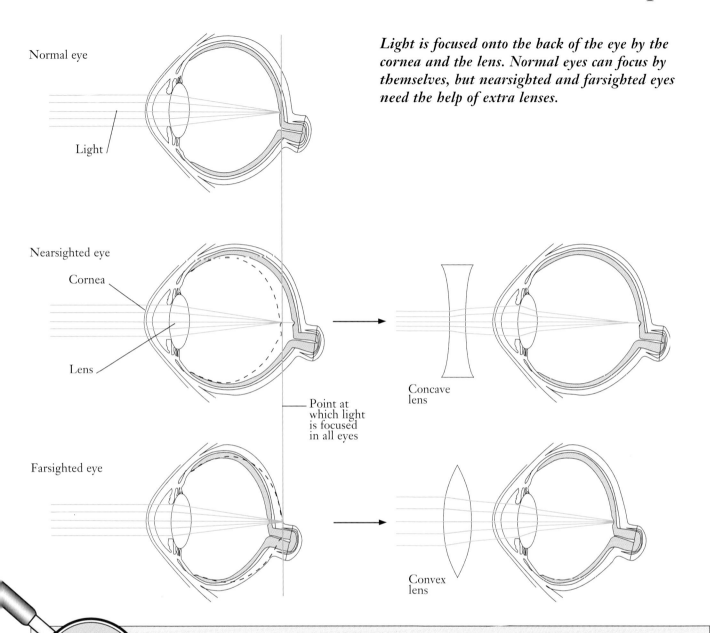

Normal eye

Light

Nearsighted eye

Cornea

Lens

Light is focused onto the back of the eye by the cornea and the lens. Normal eyes can focus by themselves, but nearsighted and farsighted eyes need the help of extra lenses.

Point at which light is focused in all eyes

Concave lens

Farsighted eye

Convex lens

The World's Most Perfect Mirror?

LOOK CLOSER

NASA's $2-billion Hubble Space Telescope contains what is probably the most perfect optical instrument. A giant mirror 8 feet (2.4 m) wide collects light from distant stars and galaxies and focuses them onto cameras and scientific instruments.

Although Hubble's mirror took years to design and polish to perfection, a faulty piece of testing equipment meant that it produced blurred pictures when it was first launched. The problem was solved by a space shuttle mission in 1993. Astronauts installed extra mirrors and lenses and ensured that Hubble was soon producing the breathtaking pictures of the Universe for which it is now known.

Light has always been an important form of communication. Visual signals, such as smoke signals or semaphore (a system of signaling by the use of two flags), relied on light to send a message from one place to another. Modern forms of communication such as fiber-optic cables also rely on light. Unlike a normal metal wire, a single fiber-optic cable can carry thousands of individual telephone conversations using pulses of laser light. Because fiber optics can carry so many signals, they have made possible exciting modern technologies such as cable television and the Internet.

Light is at the heart of some of today's most advanced technology. Lasers (high-powered, very pure sources of light) are used for everything from cutting out clothing fabrics to scanning purchases at the store checkout.

Today's computers rely on light in all sorts of ways. Tiny lasers read and store the information in CD players. Thousands of light-emitting crystals are used to make the very thin LCD (liquid-crystal display) screens used in laptop and palm-top computers. Some scientists believe that the super-fast computers of the future will use light in the same way that today's computers use electronics, because light can travel so much faster than electricity.

The story of optics

Lasers, CDs, and DVD (Digital Versatile Disc) players are today's state-of-the-art in optics and show just how far technology has advanced since scientists first began to ponder the nature of light. Greek philosopher Plato (around 428–348 B.C.E.) thought that light was a beam sent out by the eyes that made things visible when it hit them. Euclid of Alexandria (worked around 300 B.C.E.), who invented much of modern geometry, showed how light travels in straight lines and also worked out the law of reflection. Aristotle (384–322 B.C.E.), who was Plato's student, was the first person to realize that light was reflected from objects into the eye.

Many centuries later, scientists knew enough about optics to begin making instruments that could improve the seeing power of the human eye. English philosopher Roger Bacon (around 1220–1292) studied lenses and demonstrated how

Diffraction and Interference

If a girl pushes her finger into a still pond of water, ripples spread out in a circle. If she does the same thing again nearby, ripples spread out from a second circle. Eventually, the two sets of ripples meet and make a complicated pattern. This is called interference. If a boy looks at a streetlight and almost closes his eyes, he can see an unusual string of light and dark patterns. The closer his eyelids get together, the more the pattern seems to spread. This effect is called diffraction. Interference and diffraction have told optical scientists a great deal about what light is.

The fact that light will make interference patterns suggests that it is a wave. Interference explains why soap bubbles and pools of oil on the road seem to be colored. When a beam of light hits the soap or oil, part of the light is reflected back from its top surface. However, part of it passes through and reflects from the bottom surface of the film. This produces two reflected light waves that travel slightly different distances. When these waves meet, they produce an interference pattern that people see as a pattern of swirling colors.

Light travels in straight lines, but not always in the way that might be expected. It will bend around a very small obstacle, such as a sharp edge, making it look fuzzy. It will spread out as it passes through a tiny hole or slit. This happens in much the same way as sound bends around doorways, allowing people to hear around corners. Diffraction explains why there is sometimes a spot of light in the center of a shadow that forms behind an obstacle (see below).

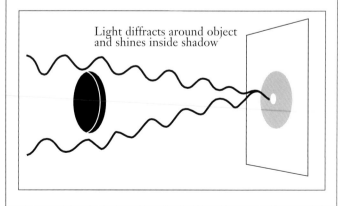

Light diffracts around object and shines inside shadow

STORY OF SCIENCE

Newton and his Reflecting Telescope

Early telescopes were a considerable improvement on the magnifying power of the naked eye. However, they suffered from the problem that different colors seemed to come to a focus at different points in the telescope, making a blurred image.

When Sir Isaac Newton shone light through a glass prism, he found it would split light into a spectrum of different colors. This led him to realize that white light is made up of different colored light rays. Furthermore, if a prism could bend the different colored rays by different amounts, the lens inside a telescope must be doing the same thing. Newton solved the blurring problem in 1668 by designing a telescope (see picture) that used a concave mirror to focus the light rays instead of a lens. This new device was called a reflecting telescope and it revolutionized astronomy. Today's largest telescopes work using concave mirrors.

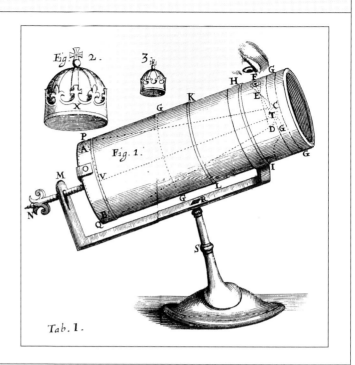

a telescope could be constructed by combining them. The name of the person who successfully constructed the first telescope is unknown. One of the earliest successful users of this new optical technology was Italian astronomer and mathematician Galileo Galilei (1564–1642), who made many discoveries with his telescope.

Around the 17th century, optics became much more of a pure science. In 1621, Dutch astronomer and mathematician Willebrord Snell (1580–1626) worked out the precise law of refraction, which is now called Snell's law. French philosopher René Descartes (1596–1650) published his version of the law in 1637, which included the observation that light travels at different speeds through different substances.

All of these scientists and many others helped to work out a general theory of how light behaves. However, it was still not clear exactly what light is. British scientist Robert Hooke (1635–1703) and Dutch physicist Christiaan Huygens (1629–1695) suggested that light is a wave. The properties of reflection, refraction, and interference could all be explained that way. The wave theory was not accepted by other scientists. English physicist Sir Isaac Newton

(1642–1727) insisted that light is a train of particles, not a wave. Not realizing how small the wavelength of visible light would turn out to be, Newton considered the sharp edges of shadows to be impossible if light were a wave.

The 19th century brought a much better understanding of light. British physicist Thomas Young (1773–1829) worked out a detailed wave theory of light and used it to explain the phenomenon of interference. Another celebrated British physicist, James Clerk Maxwell (1831–1879), showed the relationship between electricity and magnetism and discovered that light is an electromagnetic wave.

It was not until the beginning of the 20th century that physicists truly understood the nature of light. British physicists Sir James Jeans (1877–1946) and Lord Rayleigh (1842–1919) and German physicist Max Planck (1858–1947) helped to explain the idea that light could be both a wave and a particle at the same time.

CHECK THESE OUT!

✔COLOR ✔FIBER OPTICS ✔HUBBLE SPACE TELESCOPE
✔LASER ✔LIGHT ✔LIQUID CRYSTAL ✔MIRAGE
✔PHYSICS ✔QUANTUM THEORY

Ordovician Period

The geological time period that lasted from 505 to 438 million years ago

The Ordovician (OR-duh-VIH-shuhn) period is the second oldest in the Paleozoic era, following the Cambrian period. During Ordovician times, the continents looked nothing like they do today. Modern North America was turned through 90 degrees, so that the New York coastline faced south, and was joined to Greenland to make the ancient continent of Laurasia. The other main landmass was Gondwana. This was made up mainly of what is now South America, southern Europe, India, Australia, and Antarctica.

Ordovician rocks

Rocks formed in the Ordovician period are found in many parts of the world, often in mountainous areas. Mudstones and shales (fine-grained sedimentary rocks) are very common Ordovician rocks that formed in deep oceans. Volcanic rocks from this time are common in some areas. Island arcs like those that occur today in Japan are the remains of Ordovician volcanoes.

Much of North America was covered with a huge shallow sea in those times, and the climate was humid and tropical. Toward the end of the Ordovician period, there were several changes and the climate cooled. In North America, the rocks formed during the Ordovician period are divided into groups. The oldest is the so-called Ibex series of rocks, which contains limestone and sandstone. Following this are the pure quartz sandstones of the Whiterock series of rocks. At the end of Ordovician time, the Mohawk series of rocks were deposited.

Fossil starfish found in sedimentary rock that is about 450 million years old. Starfish fossils from the Ordovician period are very rare.

Ordovician fossils and life

Many life-forms developed in the warm, humid climate and extensive shallow seas. Fossils from Ordovician rocks show an abundance of brachiopods and other shelled creatures such as mollusks. One group of mollusks related to the modern nautilus and octopus, the nautiloids (NAW-tuh-loydz), swam near the bottom, grasping at food on the seabed with their tentacled arms. Corals developed during the period, coloring the shallow seabed and building reefs. Living in the sea, often near the surface, were graptolites. These delicate creatures are often found fossilized in great numbers in mudstones and shales. The fiercest animals on the sea bed were the sea scorpions (eurypterids). These giant predators grew up to 9 feet (2.7 m) long and had bodies rather like giant lobsters.

One of the most important groups during the Ordovician period was the trilobites (TRY-luh-byts). They had a shell made of segments, like that of a sow bug. Many had large eyes similar to those of a modern dragonfly.

CHECK THESE OUT!
✔CAMBRIAN PERIOD ✔GEOLOGIC TIMESCALE
✔PALEOZOIC ERA ✔SILURIAN PERIOD

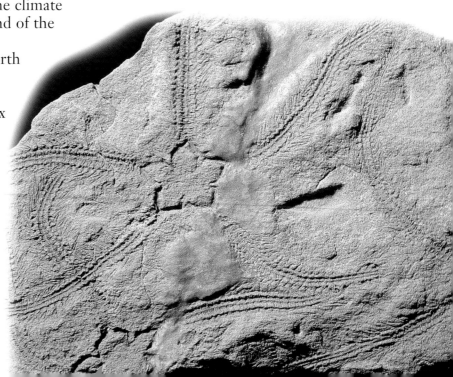

Glossary

absorb To suck or draw in. Alternatively, to receive sound or light without reflecting it or producing an echo.

bellows Instrument that draws air in and puffs it out again to provide a blast of air; used to help combustion (burning).

blizzard Long, severe snowstorm or a strong, snowy wind.

blood type Composition of a person's blood. There are four blood types: A, B, AB, and O.

by-product Something produced by an industrial or biological process; not the main product.

chemical analysis Investigating the chemical make-up of a substance.

compress To squeeze a substance so it takes up less room.

counterbalance To offset a weight with a similar weight.

culture (KUHL-chuhr) Sample of living material grown in a prepared nutrient-rich substance.

data (DAY-tuh) Group of facts used to reach a conclusion.

deflect (dih-FLEKT) To turn something off its course.

diffusion (dih-FYOO-shuhn) How particles of gas or liquid intermingle as a result of their random movment.

evaporation (ih-VA-puh-RAY-shuhn) When a liquid turns into a gas.

forecast Outline of how something is expected to happen in the future.

gas giant Four large planets in the Solar System: Jupiter, Saturn, Uranus, and Neptune. So-named because they are mostly gas.

gill Organ (as in a fish) used to remove oxygen from water.

graphite (GRA-fyt) Soft, black form of carbon that conducts electricity and is used in pencils.

hibernate (HY-buhr-NAYT) To sleep for winter while food is scarce.

industrialized world Parts of the world where more people work in industry and manufacturing than in farming. Includes Japan, Europe, and North America.

longsighted Being able to see distant objects more clearly than close-up objects.

medium (MEE-dee-um) Material through which energy—for example, sound—travels.

mollusk (MAH-luhsk) Unsegmented, spineless animal such as a snail, clam, or squid, often enclosed in a hard shell.

naked eye Unaided vision—not using a telescope or binoculars.

nautilus (NAW-tuh-luhs) Mollusks of the South Pacific and Indian Oceans with spiral shells that have pearly insides.

oasis (oh-AY-suhs) Fertile area in a desert or other barren region.

resinous (REH-zuh-nuhs) Having the properties of resin—a sticky, yellowish sap secreted by plants.

resonance (REH-zuh-nuhnts) Vibration of large amplitude (that is, with high waves); creates a rich sound.

river basin Area of land drained by a river and its tributaries (branch rivers).

rocket booster First stage of a multistage rocket that provides thrust for the launch and initial part of the flight.

shortsighted Unable to see distant objects clearly.

suffocate To stop an animal breathing by blocking its airway.

summit Highest point, for example of a mountain.

tetrahedron (TEH-truh-HEE-druhn) Three-dimensional shape that has four faces.

velocity (vuh-LAH-suh-tee) How quickly an object moves in a certain direction.

volatile (VAH-luh-tuhl) Readily vaporized (turned to gas).

Index

Page numbers in **boldface type** refer to main articles and their illustrations. Page numbers in *italic type* refer to additional illustrations.

550
EXP
#7

Exploring Earth and Space Science

05/06	DATE DUE